Environmental

PRINCIPLES AND ETHICS

Environmental

PRINCIPLES AND ETHICS

Ming H Wong
Frank W K Lee
Martin K F Fung
Hong Kong Baptist University

World Scientific
NEW JERSEY • LONDON • SINGAPORE • BEIJING • SHANGHAI • HONG KONG • TAIPEI • CHENNAI

Published by

World Scientific Publishing Co. Pte. Ltd.
5 Toh Tuck Link, Singapore 596224
USA office: 27 Warren Street, Suite 401-402, Hackensack, NJ 07601
UK office: 57 Shelton Street, Covent Garden, London WC2H 9HE

Field Trip Guide to accompany
ENVIRONMENTAL PRINCIPLES AND ETHICS

Acknowledgements

The authors would like to thank all colleagues in the Croucher Institute for Environmental Sciences and the Department of Biology, Hong Kong Baptist University especially Ms. Doris Ng for her administrative assistance. Financial support from the Teaching Development Grants (Project No.: HKBU-5 for 2001–04 Triennium) of University Grants Committee is gratefully acknowledged.

Authors

M.H. Wong, F.W.K. Lee and M.K.F. Fung

Environmental Principles and Ethics: Field Trip Guide
環境法則與倫理：野外研習指引

1. Introduction 引言 **1**

2. Tips for Organizing Field Trips 進行野外研習的注意事項 **4**

3. Field Trips in Natural Habitats 自然生境中的野外研習 **7**

3.1 Wetlands 濕地 **8**

3.2 Woodlands 林地 **21**

3.3 Streams and Shorelines 河流及海岸 **34**

4. Field Trips Applying Environmental Principles 應用環境法則的野外研習 **49**

4.1 Applying Environmental Awareness 應用環保意識 **51**

4.2 Applying Corporate Environmental Management 應用企業環境管理 **58**

4.3 Applying Sustainable Development 應用可持續發展 **65**

4.4 Applying Resources and Waste Management 應用資源與廢物管理 **72**

4.5 Applying Food Safety 應用食物安全 **79**

4.6 Applying Organic Farming 應用有機耕作 **89**

4.7 Applying Ecotourism 應用生態旅遊 **96**

Reading Materials and Contacts 讀本與聯絡方法 **103**

Index **105**

Note 記事 **107**

1. Introduction

About Croucher Institute for Environmental Sciences

This publication (along with the education package entitled Environmental Principles and Ethics) has been proposed and implemented by the Croucher Institute for Environmental Sciences (CIES), Hong Kong Baptist University.

Hong Kong Baptist University

Led by **Prof. Ming Hung Wong**, (Chair Professor of the Biology Department of Hong Kong Baptist University and Director of CIES), CIES has been involved in research and consultancy projects related to environmental impact assessment, revegetation of derelict land, and treatment and recycling of industrial, agricultural and domestic waste, with emphasis on the use of relatively low cost technology. Through its many distinguished research projects and consultancy services, CIES has, over the years, acquired an international reputation and established international, national and local links.

Since April 2004, the Croucher Institute for Environmental Sciences has been named after the **Croucher Foundation** as the only awarded institution within the Hong Kong Baptist University. CIES was formerly known as the Institute for Natural Resources and Environmental Management (INREM, 1997-2004) and the Centre for Waste Recycling and Environmental Biotechnology (1990-1997) which was inaugurated in 1990. The honourable naming by the Croucher Foundation has allowed the Institute to strengthen in our existing research and consultancy services.

For more information about CIES, please visit www.cieshk.org.

Importance of Field Trips in Environmental Education

Typical image of Hong Kong

Despite its small size of 1,000 sq kilometers and its bustling international city image, Hong Kong is teeming with natural environment. Hong Kong has a wide range of habitats, including wetlands, woodlands, streams, open country, farmland, shrubland, grassland, coastal areas and urban areas.

This publication aims to make use of natural environment and other related premises as outdoor classrooms to illustrate the importance of conservation and environmental principles.

Not only there is no limitation from textbook and teaching formula, the "nature" classroom is full of teaching materials everywhere. Using real species and practical objects rather than artifacts or specimens is the best way to arouse students' interest. Let them listen to the sounds of nature, and they will be attracted to birds singing on the mangrove trees. Give them identification key, and let them use their strong investigation skills and search for insect species on the grass. Students need not only knowledge but also a chance to go into the natural world. After all, the relationship between human and nature is basic knowledge for the students.

Beautiful scenery at Mai Po

Tai Ho River in Lantau Island

EPD Resource Centre in Fanling

The first part of this book focuses on **field visits in natural habitats**, namely:

- wetlands
- woodlands
- streams and shorelines

These sites offer direct learning experiences for the students to interact with physical and biotic environment of nature.

The second part of this book deals with **field visits applying environmental principles**, such as EPD Environmental Resource Centres, Hong Kong Planning and Infrastructure Exhibition Gallery, Fung Yuen Butterfly Reserve, etc. These sites are categorized by the environmental principles they apply to. By visiting these atypical outdoor classrooms, students can practice different environmental principles taught by the textbook and CD-Rom in the education packages.

This book could be used as a reference material by academic institutions, non-profit organizations and government departments. It provides practical educational materials on the topic of life science and ecotourism, which aids in increasing the knowledge on the environment.

1. 引言

關於裘槎環科所

本刊物（連同名為「**環境法則與倫理**」的教材套）由香港浸會大學‧裘槎環科所的員工籌備及製作。

由香港浸會大學生物系**黃銘洪講座教授**帶領，裘槎環科所是集科研、教育、學術交流、資訊傳遞和人才培訓於一體的多功能環境科學中心，竭誠為香港、珠江三角洲乃至整個東南亞地區提供嶄新的資源管理理念和環境諮詢服務。其所涉及的研究和諮詢領域涵蓋空氣、土壤、水、生物等自然資源以及各種無機和有機污染物的環境行為及其風險評價。

自二○○四年四月起，本所有幸被**裘槎基金會**冠名為「裘槎環科所」，作為香港浸會大學唯一獲此殊榮的學術研究所。是次冠名除對本所以往工作成就的高度肯定和認同外，還大大加強了本所現有或將來的研究項目及諮詢服務。

詳情請瀏覽網址：www.cieshk.org

野外研習在環境教育的重要性

香港雖為寸金尺土的國際都會，但仍富有不少珍貴的天然美景，不論是濕地、木林、河溪、郊區、農地、叢林、草地、海岸或市區，全都是獨特的生態環境。

本刊物旨在利用自然環境及其他相關場地，讓學生作戶外教室，學習自然保育和環境法則的重要性。野外研習不但沒有像書本的學習模式的規範，而且當置身戶外教室，學生更可以隨處就地取材，活學活用。無論是紅樹上在歌唱中的雀鳥，或是在草地上的昆蟲品種，各式各樣的生物和現場實物活生生呈現於學生眼前，統統都是實用的教學材料。畢竟野外研習是最有效的學習方式！

本書的第一部分著重推廣**自然生境中的野外研習**，題目包括：濕地、林地、河溪和海岸。這些地點供學生直接學習了解大自然和生態環境的機會。

第二部分則介紹一些**應用環境法則的野外研習**，例如參觀環境保護署環境資源中心、香港規劃及基建展覽館、大埔鳳園蝴蝶保育區等，並嘗試以不同環境法則作分類。透過參觀這些「另類」的戶外教室，學生可以把課本和光碟中所學到的環境知識，融會貫通，應用於日常生活中。

本刊物可供本港各大學、非牟利團體及政府部門作參考材料，當中包含有關生物學和生態旅遊的實用教學材料，有助加深對自然環境的認識。

2. Tips for Organizing Field Trips

Field trips could be joyful learning experiences when detailed planning is prepared and study guidelines are followed. Like all outdoor activities, no matter where you visit, safety always comes first!

In General:

- Carefully plan and familiarize yourself with the route before setting out. Escape routes and contingency plans should be arranged in advance.
- Note the weather conditions, pay attention to potential climatic hazards.
- Put on suitable clothes and footwear.
- Take plenty of water and drink frequently to prevent dehydration.
- For emergency, bring a first-aid kit.
- Inform someone where you are going.
- Do not litter.
- Do not pollute the local environment.

Plan ahead carefully

During site visits, students are encouraged to appreciate the rich and diverse species and remind not to disturb their habitats. It is important to protect our natural environment and wildlife by the following guidelines.

For Field Trips in Natural Habitats:

- Do not hurt or disturb any wildlife. It is illegal to remove any animals or plants from Country Parks or Special Areas in Hong Kong.
- Some animals (and plants) can be dangerous especially snakes and the nests of bees and wasps, so leave them alone and warn others of their presence as well.
- Wear protective clothes and footwear in outdoor environment.
- Use sun block to protect against sunburn.
- Do not eat wild fruit or drink untreated water from the streams.
- Do not smoke or start fires at any time.
- Follow the specific guidelines in each respective chapter in **Part 3**.
- Remember the motto of ecotourism: "**take nothing but photos, leave nothing but footprints**".

Don't disturb wildlife!

Unlike natural habitats, most of the field trips applying environmental principles take place inside special premises. Although these sites are not affected by physical factors such as weather conditions, students should still be cautious and follow the respective guidelines when visiting these venues.

Take safety precautions if needed

For Field Trips Applying Environmental Principles:

- Make appointment with local management authority before the visit. Pay attention to the opening hours of the premises.
- Carefully plan and familiarize yourself with the route before setting out. Beware of safety inside the premises.
- Listen and follow the advice of local management authority. Do not disturb the local operations.
- Don't take photos unless approval is granted from local management authority.
- Follow the specific guidelines in each respective chapter in **Part 4**.

Enjoy the happy learning experience!

2. 進行野外研習的注意事項

如果有詳盡的計劃和跟隨指引，野外研習必會是愉快的學習體驗。像所有戶外活動一樣，無論你參觀哪裡，首先要注意安全！

一般事項：

- 在出發前，小心計劃和熟悉路線，並且安排撤離路線和應急計劃。
- 留意天氣狀況及氣候變化。
- 穿上合適的衣服和鞋類。
- 攜帶大量清水和經常補充水份，以防止脫水。
- 攜帶急救工具以應付緊急情況。
- 告知他人你前往的地方。
- 不要亂拋垃圾。
- 不要污染環境。

考察其間，鼓勵學生欣賞物種的豐富和多樣性，以及提醒他們不要打擾物種的棲息地。根據以下指引，保護我們的自然環境和野生動物。

自然生境中的野外研習：

- 不要傷害或打擾任何野生動物。在郊野公園或特別地區採集動物或植物，實屬違法。
- 一些動物（和植物）尤其是蛇，蜜蜂和黃蜂的巢穴具危險性，因此別打擾它們，並警告其他人。
- 在戶外環境，穿戴保護衣服和鞋類。
- 使用防曬用品，保護皮膚。
- 不吃野生水果或飲用未經處理的溪水。
- 任何時候都不可以吸煙或生火。
- 跟隨**第三部份**的個別指引。
- 緊記生態旅遊的格言：「**不要取走任何東西，除了照片；不要留下任何東西，除了足印。**」

與自然生境不同，大部分應用環境原則的考察都是在指定的室內進行。儘管這些地點不受天氣情況等因素影響，不過學生都應該謹慎和跟隨指引。

應用環境法則的野外研習：

- 在參觀前，跟管理組織預約。留意開放時間。
- 在出發前，小心計劃和熟悉路線。留意參觀地點的安全。
- 聽從管理組織的忠告。不要打擾組織的運作。
- 除非得到批准，否則不要拍照。
- 跟隨**第四部分**的個別指引。

3. Field Trips in Natural Habitats 自然生境中的野外研習

Hong Kong is known for its cityscape and highly urbanized metropolis. It is much less well known for its unusually high and extremely diversified habitats and wildlife. Here we have mountains, streams, oceans, forests, mangroves, as well as precious animals such as Chinese White Dolphins, Hong Kong Newt, etc. Recent surveys indicate that there have been further increases in the numbers of some local species: 110 dragonflies, 260 butterflies (one sixth of the total butterfly species in China) and 452 birds (one third of total bird species in China).

Among all the existing natural habitats found in Hong Kong, this Field Trip Guide focuses on field visits in:

1) **Wetlands**
2) **Woodlands**
3) **Streams and Shorelines**

These sites offer direct learning experiences for the students to interact with physical and biotic environment of nature.

香港以其城市景觀和都市發展嚮譽全世界，不過較少人會留意到，這個面積細小的大都會其實擁有豐富的生態環境及極多樣性的生物品種。事實上，我們四周還有山脈、河流、海洋、森林、紅樹林，以及珍貴物種如中華白海豚、香港蠑螈等。根據最近的調查顯示，本地品種的數量正不斷上升：約110種蜻蜓、260種蝴蝶(佔全中國蝴蝶品種的六分之一)、452種鳥類(佔全中國鳥類品種的三分之一)。

在本港眾多自然生境中，本野外研習指引現介紹以下地點：

1) 濕地
2) 林地
3) 河流及海岸

這些地點供學生直接學習了解大自然和生態環境的機會。

3.1 Wetlands

Precious wetlands in Mai Po

Wetlands are precious ecological habitats, where water serves as the main control of the environment. At low tide, wetlands include areas of marine water where the depth could be as low as six metres. They cover coastal habitats including coral reefs, mudflat and mangrove forests. According to the nature of the water sources, wetlands can be classified as marine, estuarine, riverine, palustrine and artificial.

As "cradles" of biological diversity, wetlands host many plant and animal genetic resources. They supply water and food nutrients on which many species of plants and animals depend. Wetlands also function as stream and flood control, groundwater replenishment, water purification, sediment and nutrient retention, climate change mitigation, and shoreline stabilization and protection.

In addition, wetlands are breeding and nursery grounds for fish (i.e. *gei wai* farming), and for many other aquatic or marine crops for human use, bringing significant economic benefits. Thus, they enhance the health and welfare of the local people. Besides, wetlands also contribute to recreation, education and ecotourism by enriching the scenic beauty and strengthening the traditional and cultural heritage.

gei wai farm

In Hong Kong, the varieties of wetland are often associated with inter-tidal mudflats, fishponds, sea-grass beds, coral reefs, rocky shores, rivers and streams, freshwater marshes, irrigated agricultural land and reservoirs.

Key Species in Wetlands

Wetland is well-known for its rich diversity of animals and plants. Undoubtedly, the migratory birds and mangroves found in wetlands are of great ecological value. Besides, there are also many important fauna and flora such as aquatic plants, mammals, reptiles, amphibians, freshwater fish, butterflies and moths, dragonflies, etc.

Vegetation

As tropical evergreen plants at inter-tidal shores, **mangroves** have specialized adaptations to the environment in physiology and morphology. They stabilize and protect the shoreline, purify water and supply food and nutrients.

Kandelia and its droppers

Among the mangroves found in Hong Kong, *Kandelia obovata* is the known for its abundance in various mangals and for the special shapes of the droppers. Other mangrove species of Hong Kong include Black Mangrove (*Avicennia marina*), *Aegiceras corniculatum*, Many-petaled Mangrove (*Bruguiera gymnorrhiza*), Blind-your-eye (*Excoecaria agallocha*), and Spiny Bears Breech (*Acanthus ilicifolius*).

Many-petaled Mangrove

Grasses and **sedges** are quite common in wetlands, too. Most *gei wais* have been densely occupied by species such as Common Reedgrass (*Phragmites australis*), a reed featuring long, slender leaves and a plume-like inflorescence. Short-leaved Malacea Galingale (*Cyperus malaccensis*) and Coastal Bulrush (*Scirpus littoralis*) survive in water with low salinity.

Blind-your-eye

Common Reedgrass

Little Egret

Ducks

Wetland Birds

In Hong Kong, most of the 450 bird species are winter visitors and passage migrants. They can be found in freshwater wetlands like fishponds where fishermen pump water out of the fishponds during the winter. The trash fish left in the small pools attracts large numbers of **egrets** to feed. The muddy bottom also provides nursery grounds for **waders**.

Egrets, herons and bitterns are waterbirds with long legs, long neck and long bills. Species like **Little Egret** (*Egretta garzetta*), **Great Egret** (*Egretta alba*), **Chinese Pond Heron** (*Ardeola bacchus*) can be found nesting in trees and feeding their juveniles during the breeding season.

Classified as endangered species in 2001, **spoonbills** are characterized by long legs and a long spoon-shaped bill. Among the six species of spoonbills in the world, two of them, **Black-faced Spoonbill** (*Platalea minor*, 22% of world population) and **Eurasian Spoonbill** (*Platalea leucorodia*) are local visitors. They migrate to wetlands and coastal areas in Taiwan, Hong Kong, Northeastern China, Japan and Vietnam.

Shorebird

Belonging to Family Anatidae, **ducks**, **geese** and **swans** are waterbirds with a flattened bill, short body, a short tail and legs and webbed feet. Other common waterbirds include:

- **kingfishers** (e.g Common Kingfisher)
- **shorebirds** (e.g. plovers, sandpipers, snipes, curlews)
- **gulls** (e.g. Black-headed Gull and Saunder's Gull) and
- **raptors** (e.g. Black Kite, eagles and falcons)

Other Animals

Freshwater fish such as **Carps**, **Eels** and **Mullets** are aquaculture species cultivated in the fishponds. In the inter-tidal mudflat lies **Fiddler Crab** (*Uca arcuata)*, **Common Mudskipper** (*Periophthalmus pectinirostris*) and **Bluespotted Mudskipper** (*Boleophthalmus pectinirostris*).

Common mudskipper

Numerous species of amphibians and reptiles live in the wetlands in Hong Kong. Some of the common amphibians in freshwater ponds are **Gunther's frog** (*Rana guentheri*) and **Pigmy Frogs**. One should be aware of the snakes such as **Indo-Chinese Rat Snake** (*Ptyas korros*) and **Mangrove Water Snake** (*Enhydris bennettii*) hiding inside the tall grass or the marshes.

Fiddler Crab

Mangroves and reed provide adequate larval food sources for many **butterfly**, **moth** and **dragonfly** species. The common butterflies and moths include Common Black Jezebel (*Delias pasithoe*) and Pneumatophore moth (*Hymenoptychis sordida*). The endangered dragonfly, Four-spot Midget (*Mortonagrion hirosei*) has been recorded in Mai Po.

Butterfly

Although many mammals are elusive and difficult to spot, visitors may still find some small species such as **Chinese Otter** (*Lutra lutra chinensis*), **Javan Mongoose** (*Herpestes javanicus*) and **Chinese Leopard Cat** in the wetlands.

Wetland Conservation

Wetlands surviving in the city

Because of their productive environments, with valuable ecological, economic and scenic benefits, wetland's conservation is vital. It largely depends on the efforts of individual governments and local stakeholders. Global cooperation is also required to protect the wetlands, since they are key habitats for migratory fauna that migrate across national borders.

Established in 1971, the ***Ramsar Convention*** is an international treaty aiming to protect the national and international aspect of wetlands. It provides the framework for actions to protect conservation and appropriate use of wetlands. As of 2004, over 1,300 wetlands covering some 120 million hectares have been designated as Ramsar's "List of Wetlands of International Importance". Hong Kong's Mai Po Inner Deep Bay is one of the 21 Ramsar wetlands in China.

Futian Nature Reserve in China

Entrance of Wetland Park

Exhibits

Visitors

Hong Kong Wetland Park

As the first nature-based theme park established and operated by the Government, the Hong Kong Wetland Park is located in Tin Shui Wai in the New Territories. It provides a place to explore outdoor education on wetland conservation for the public using dynamic displays, audio-visual shows, interactive computer devices, and wildlife models. Guided tours are arranged for students and public visitors, aiming to increase their interest in understanding wetland ecosystem.

After 2005, Phase Two of the Hong Kong Wetland Park will be open. This extension will hold a 60-hectare **Wetland Reserve** consisting of various habitats such as freshwater marshes, woodlands, grasslands, reed beds, mangrove and inter-tidal mudflats. The new facilities will allow visitors to explore more on natural wetlands and migratory birds.

Website: www.afcd.gov.hk/wetlandpark

Suggested Wetland Field Sites in Hong Kong

Most of wetlands in Hong Kong are located in the northwest and east coasts of the New Territories. In terms of area size, the major ones include Mai Po Nature Reserve, Nam Seng Wai, Kei Ling Ha and Tsim Bei Tsui. Others include Tai O, San Tau and Shui Hau. This chapter explores more on **Mai Po Nature Reserve**, **Lai Chi Wo** and **To Kwa Peng**.

1) Mai Po Nature Reserve

In 1995, the Hong Kong Government listed a 1,500-ha area of the Deep Bay wetlands in the northwestern New Territories under the Ramsar Convention. As the sixth largest mangrove stands in China, the area has discovered more than 20 animal and plant species new to science.

The commercial fishponds within the Ramsar Site were ecologically important by indirectly providing feeding habitats for birds. In winter, when the fishponds are filled with harvesting fishes such as Mullets and Carps, waterbirds especially egrets, herons and the "Flagship" Black-faced spoonbills (*Platalea minor*) are attracted to these ponds to feed on the non-commercial fishes (like *Gambusia affinis*) and shrimps (like *Macrobrachium nipponense*) trapped in the shallow water.

Mai Po Nature Reserve is an important habitat for over 10,000 migratory birds every year. Over 250 species have been recorded here, and more than half this number are migrants from December to February. The key species include Egrets (site holds about 30% of Hong Kong), Pied Avocets (about 6,000 recorded in 2003), Dalmatian Pelicans and Saunder's Gulls (both are endangered species), Black-faced Spoonbills (one-quarter of the world population winters recorded and about 300 individuals spotted in the winter of 2004-05), Waders and Shovelers.

Entry to Mai Po Nature Reserve is restricted under the Wild Animals Protection Ordinance. Individual or groups can join the guided ecotours offered by the World Wide Fund for Nature Hong Kong. During the tour, visitors may watch birds from hides; walk trails by fishponds and traditional shrimp ponds (*gei wais*); and head along a floating boardwalk that leads between mangroves to a hide on the edge of the Deep Bay mudflats bordering Hong Kong and mainland China.

Transportation: Take the 76K Bus from Sheung Shui to Yuen Long. Follow Tam Kon Chau Road at the left side of the Mui Sze Store and take a 20-minute walk to the Mai Po Visitor Centre.

Suggested Time: 4 hours at the Reserve

Entrance

Walking through mudflats

Bird-watching hide

Mudflat in the Deep Bay

Guided tour

2) Lai Chi Wo

Overview

Heritiera forest

Heritiera fruit

Mudflat

Lai Chi Wo village and *feng shui* woodland

Located at the northeastern shoreline of the New Territories, Lai Chi Wo embraces challenging landscape, beautiful shores as well as abundant biodiversity. Perhaps the only drawback of this site is its extremely remote location: it takes 2-3 hours of walking distance from the nearest transportation point at Luk Keng or Wu Kau Tang.

In Lai Chi Wo, an uncommon mangrove species called **Coastal Heritiera** (*Heritiera littoralis*) has formed the largest and most primitive forest. Many trees are so large that it takes two people to reach around their trunks, while some buttress roots grow to as tall as 1 metre.

Outside the Coastal Heritiera forest, a large mudflat features great biodiversity and dynamics. Mangroves live in the anaerobic wet mud through unusually shaped roots. Some of them include the Black Mangrove (*Avicennia marina*), the Many-petalled Mangrove (*Brugiera gymnorrhiza*), and Lumnitzera (*Lumnitzera racemosa*).

The mudflat also hosts an important sea-grass called Dwarf Eel Grass (*Zostera japonica*), a flowering plant with long narrow leaves which can be completely submerged in seawater. As a key species in the coastal ecosystem, sea-grass provides food and shelter for many young coastal creatures, such as clithons, transparent fish fry and shrimp larvae. In summer, animals such as insects, butterflies, damselflies and water buffaloes all become energetic in the neighbourhood.

It is easy to find traditional village culture in Lai Chi Wo. The village comprises numerous farmhouses arranged in neat rows with a large piazza and a few temples. This characteristic hamlet symbolizes the Hong Kong's traditional, rural communities.

Transportation: Take minibus (56K) from Fanling KCR station to Luk Keng. Then walk from Luk Keng along the shoreline and the country roads towards Lai Chi Wo.

Suggested Time: 3 hours at the site.

3) To Kwa Peng

On the east side of Hong Kong in Sai Kung, To Kwa Peng is another good example of unique wetland environments and diverse ecosystems. In just about 20 minutes' walk from the Uk Tau bus station, one can arrive to the bay full of mangroves.

Overview of To Kwa Peng

The uniform mangrove bed nourishes many species of different distinctiveness: *Aegicaras corniculatum* with salt-laden leaves and bright red petioles, Black Mangrove (*Avicennia marina*) with opposite leaves with grey undersides, and *Kandelia obovata* with slender green droppers in spring season, etc.

Aegicaras corniculatum

In the coastal area is a dense woodland of Cuban Bast (*Hibiscus tiliaceus*) and Chinese Hibiscus (*Hibiscus rosa-sinensis*). With heart-shaped leaves and overlapping yellow petals, Cuban Bast is famous for its tolerance to salinity and strong wind.

A little pier is located outside the Cuban Bast forest. The nearby mudflat is occupied by mudskippers. Small crabs and periwinkles (*Littoraria spp.*) are also found on mangrove leaves and stems.

Littoraria spp.

Transportation: Take Bus No. 94 from Sai Kung (or 96R from Diamond Hill every Sunday and public holiday) towards Wong Shek Pier. Get off at Uk Tau, cross the road and follow the footpath to To Kwa Peng.

Suggested Time: 2 hours at the site

General Guidelines for Wetland Field Trips

Please respect wetlands!

- Beware of daily tidal changes.
- Observe wetlands quietly.
- Do not hunt, disturb or collect any wildlife.
- Do not step on sea grasses, and roots or seedlings of mangroves.
- Take away your rubbish.
- Respect village traditions, treasure villagers' properties.
- Please keep a distance from wild birds and do not disturb them or their eggs and nests.

Practical Worksheet for Wetland Field Trips

1) Area of site visited: ______________ hectares

2) Name 4 animal species that you found in the site:

___________________ Distinctive features: ________________________

___________________ Distinctive features: ________________________

___________________ Distinctive features: ________________________

___________________ Distinctive features: ________________________

3) Name 3 mangrove plants that you found in the site:

___________________ Distinctive features: ________________________

___________________ Distinctive features: ________________________

___________________ Distinctive features: ________________________

4) Problems encountered by mangroves in the site:

____

____

5) How can mangroves adapt to the tough conditions of wetland?

____

____

6) Please suggest some of the ways to conserve wetland:

____

____

3.1 濕地

濕地是賦有自然生態價值的獨特環境，這些地區的環境由水主宰。濕地既有天然的，也有人工建造的，有屬永久或臨時的。濕地的水體可以是靜止的，有的則為流動的；可以是淡水、鹹水或鹹淡水。濕地也涵蓋部分潮退時水深不超過6米的海水水域。

雖然濕地的特性是受到水的影響，但因應不同的環境因素如高度、氣候、地點及鹽濃度，世界各地形成了多樣化的濕地類別。根據不同的水源性質，濕地可分為6大類 — 海洋、河口三角洲、湖泊(包括與湖泊有關連的濕地)、河濱(包括與河溪相關連的濕地)、淡水(包括位於沼澤、草澤及苔地的濕地)和人工濕地。我們可在香港找到不同類別的濕地，包括潮澗泥灘、魚塘、海草床、珊瑚礁、岩岸、河流和水溪、淡水沼澤、灌溉農田和水庫等。

由於濕地擁有充沛的水源、肥沃的土地、有機物和其他自然資源，因此早期十分吸引人們前來定居。事實上，在世界各地和人類歷史上，濕地一直是不同文化的發源地，亦為人們提供生活所需。雖然如此，濕地近年常被誤以為是水浸的源頭，或是影響人類健康的昆蟲和其他害蟲的繁殖地，每當由昆蟲傳播的傳染病爆發，濕地往往首先被認為是要清除和消毒的地方。

濕地功能包括：

- 自然保育
- 科學研究
- 環境教育
- 農業
- 基圍蝦塘
- 休憩
- 農業

濕地的主要物種

濕地以擁有豐富而多樣化的動植物見稱，遷徙鳥和紅樹林無疑都具有重要的生態價值，此外，這裏亦孕育著其他重要的動植物如水生植物、哺乳類、爬行類、兩棲類、淡水魚類、蝴蝶、蛾及蜻蜓等等。

紅樹及其他植物

紅樹是一種常綠植物，只生長在熱帶和亞熱帶的潮間帶岸邊，具有獨特的生理結構及形態去適應環境。紅樹的功用包括穩定及保護海岸線、淨化水質及提供食物和養料。

秋茄 (*Kandelia obovata)*是香港常見的紅樹品種之一，其繁殖體外形獨特，本港其他紅樹品種包括白骨壤 (*Avicennia marina*)、桐花樹 (*Aegiceras corniculatum)*、木欖 (*Bruguiera gymnorrhiza*) 、海漆 (*Excoecaria agallocha*) 及 (*Acanthus ilicifolius*)。

濕地鳥類

香港現時有450種鳥類，大部分為冬候鳥和遷徙鳥，常見於淡水濕地如魚塘，漁民會於冬天時將水塘抽乾，剩下一些沒有經濟價值的魚，因而吸引大量的**白鷺**前來覓食，而魚塘的泥灘亦會吸引大量**涉禽**。

鷺和鳽是鷺科水鳥，特徵是有一雙長腿、長頸和長嘴。有些鷺鳥品種如小白鷺 (*Egretta garzetta*)、大白鷺 (*Egretta alba*) 和池鷺 (*Ardeola bacchus*) 每逢繁殖季節都會在樹上築巢，養育小鷺。

琵鷺擁有長腳及琵琶狀的長咀。全世界共有六個琵鷺品種，香港共錄得其中兩種，分別是黑臉琵鷺 (*Platalea minor*, 佔世界總數的百分之二十二) 和白琵鷺。根據研究資料顯示，黑臉琵鷺遷飛往台灣、香港、中國東北部、日本和越南等地區的濕地和沿海地區。

鴨、鵝和雁屬於雁鴨科水鳥，嘴巴扁平，身體短胖，尾巴和腿短小，腳趾長有蹼，便於游泳。翠鳥長有引人注目的羽毛，喜歡站在架空電線或樹枝上，瞄準目標便俯衝而下捕食小魚、小螃蟹和其他細小的水生無脊椎動物。

卷羽鵜鶘 (*Pelecanus crispus*) 為香港最大型的鳥，並列為瀕危品種，於隆冬在內后海灣出現。其他水鳥包括遷飛涉禽 (如千鳥、鷸和杓鷸)、鷗 (如紅嘴鷗和黑嘴鷗) 及猛禽 (如麻鷹、鵰和遊隼) 等等。

其他動物

池塘飼養的淡水魚品種包括鯉魚、鱔及烏頭等。潮潤泥灘可找到招潮蟹 (*Uca arcuata*)、彈塗魚 (*Periophthalmus pectinirostris*) 和大彈塗魚 (*Boleophthalmus pectinirostris*)。

超過30種兩棲類動物和爬蟲類在香港濕地棲息，當中包括沼蛙 (*Rana guentheri*) 和紋姬蛙 (*Microhyla ornate*)。此外更要留心藏身於草叢或沼澤地的灰鼠蛇 (*Ptyas korros*) 及黑斑水蛇 (*Enhydris bennettii*)。

紅樹林和蘆葦為蝴蝶、蛾和蜻蜓提供足夠的幼蟲食料。較常見的有報喜斑粉蝶 (*Delias pasithoe*) 及 *Hymenoptychis sordida* 蛾種。在米埔濕地也曾發現被列為瀕危物種的廣瀨妹蟌蜻蜓 (*Mortonagrion hirosei*)。

雖然許多哺乳類動物都是難以捉摸及記錄，遊客仍可在濕地找到一些細小品種如水獺 (*Lutra lutra chinensis*)、紅頰獴 (*Herpestes javanicus*) 和豹貓。

濕地保育

濕地是極具生產效益的環境，不論在生態和經濟上的功能都為我們帶來重大的裨益，因此必須悉心保護。

濕地是生物多樣化的「搖籃」，孕育著珍貴的動植物自然遺產，為許多動植物品種提供必須的水生環境，令自然生產循環不息，讓許不同的動植物得以繁衍。此外，濕地還有河溪防洪、補給地下水、穩定和保護海岸線、保留養分和沉積土、紓緩氣候變化及淨化水質等功用。

再者，濕地亦為人類帶來不少經濟利益，因為這裡是魚類及其他水產或海產生物的繁殖和成長基地。濕地不單促附近居民的健康和福祉，更提供了康樂、教育和生態旅遊的好去處。濕地使我們的效野更添美態，有助促進農業發展，還保存了悠久而寶貴的文他傳統。

香港濕地公園

香港濕地公園位於天水圍，毗鄰南生圍，是香港首個以自然為主的主題公園，設有充滿動感的展品、影音播放、互動的電腦裝置及野生生物模型，為市民提供一個集保育、教育及探索濕地的地方。公園還會為特定團體如學校、非謀利機構及其他公眾人士安排導覽參觀，以引起大眾對濕地生態系統的興趣。

香港濕地公園第二期將於2005年年低開幕，其中包括60公頃的濕地保護區，這裡可找到淡水沼澤、林地、草地、蘆葦叢、紅樹林和潮澗泥灘等不同的生境。新建的設施可讓遊客探索自然濕地和遷徙雀鳥。
參考網址：www.afcd.gov.hk/wetlandpark

香港濕地考察地點

香港有多片濕地，大多位於新界沿岸的西北部和東部。以面積計算，主要的濕地包括米埔自然保育區、南生圍、企嶺下、鹿頸和尖鼻咀。其他熱門的濕地包括大澳、大嶼山的水口和港島的大潭督。我們建議到米埔自然保育區、荔枝窩和土瓜坪考察。

1) 米埔自然護理區

1995年，香港政府將位於新界西北部，總面積達1,500公頃的米埔內后海灣地區正式根據《拉姆薩爾公約》列為國際濕地。米埔內后海灣拉姆薩爾濕地的魚塘佔地467公頃。這片紅樹林於中國排名第6位，曾發現超過20種科學界的動植物新品種。

早於1990年代，已有研究指出米埔的魚塘擁有極高的生態價值，能間接為鳥類提供覓食的生境。魚塘於冬天會放乾，以便捕撈塘內的魚類如灰鯔和鯉魚，而一些非商業用途的魚 (如 *Gambusia affinis*) 和蝦 (如 *Macrobrachium nipponense*) 被困於淺水區，吸引了數以百計的水鳥，特別是蒼鷺、白鷺和瀕危鳥種黑臉琵鷺 (*Platalea minor*) 飛來覓食。

米埔自然保護區為超過10,000隻每年途經該區的遷徙雀鳥提供重要的生境，這裏共錄得超過250種鳥類，其中超過一半於十二月至二月期間遷徙。主要品種包括Egrets (佔全港百分之三十)、Pied Avocets (於2003年錄約6,000隻)、Dalmatian Pelicans and Saunder's Gulls (兩者均屬瀕危品種)、黑臉琵鷺 (全球有四分一於米埔越冬，2004–05年更錄得超過300隻)、Waders and Shovelers.

這片濕地乃受《野生動物保護條例》所保護的限制地區。觀鳥人士可向漁農自然護理署申請許可証內進參觀。市民亦可參加世界自然(香港)基金會安排的生態導賞團，遊人可在位於沼澤的觀鳥屋內欣賞雀鳥，沿魚塘和傳統養蝦池(基圍)步行參觀；亦可經兩旁長滿紅樹的浮橋前往介乎香港和大陸的後海灣泥灘。

交通：乘搭由上水至元朗的76K巴士，沿Mui Sze Store左邊的担竿路步行20分鐘便到達。

建議時間：約四小時

2) 荔枝窩

荔枝窩位於新界東北沿岸，擁有嘆為觀止的景觀、迷人的海岸及豐富的生物多樣性。但這裏位置偏遠，從最近的鹿頸或烏郊騰出發，均需步行2–3小時方可到達。

銀葉樹 (*Heritiera littoralis*) 是一種罕見的紅樹品種，但在荔枝窩卻可找到最大片及最原始的銀葉樹林。當中許多樹身都粗大得要兩人才可合抱，很多板根的高度更幾可及胸。

銀葉樹林外的泥灘展現了多采多姿的面貌及旺盛的生命力，你可以看見各種紅樹如海欖雌 (*Avicennia marina*)、木欖 (*Brugiera gymnorrhiza*) 和欖李 (*Lumnitzera racemosa*) 為了讓根部在厭氧的濕泥中仍可保持呼吸，衍生出呼吸根來。

泥灘亦有著一片重要的海草床，矮大葉藻 (*Zostera japonica*)，葉長條形，可完全淹沒在海水中仍能生存的有花植物。海草在沿岸生態扮演重要角色，為許多海邊生物幼體提供哺育場所，彩螺、身體晶螢通透的魚苗和蝦仔。在蟬聲聒耳的夏天，你總可在濃密的草叢間看到輕輕飛舞，色彩迷人的豆娘，在田野上悠閒地啃草的黃牛和水牛，以及常與水牛為伴的牛背鷺 (*Bubulcus ibis*)。

你更可在荔枝窩緬懷村中往昔，站在村前的廣闊空地，看著一排排錯落有致的村屋，令人緬懷昔日村民簡樸而自給自足的鄉郊生活。

交通：前往粉嶺火車站乘56K小巴往鹿頸。由鹿頸沿岸及郊遊徑往荔枝窩。

建議時間：約三小時

3) 土瓜坪

位於西貢土瓜坪的海濱，便是香港東面一處可以讓人飽覽濕地世界裏繁複生命景致的海岸環境。從屋頭巴士站步行約12分鐘，便可到達紅樹林海灣。

這片紅樹叢由不同品種，特徵各異的紅樹所形成：葉柄通紅、葉面常滿布鹽粒的蠟燭果 (*Aegicaras corniculatum*)；葉片對生、葉底灰白的海欖雌 (*Avicennia marina*)，以及於春天掛著一條條綠色筆狀的種子胚軸的秋茄 (*Kandelia obovata*)。

小徑前方是茂密的黃槿 (*Hibiscus tiliaceus*) 樹林，葉片心形，花瓣黃色並互相重疊，黃槿十分耐鹽，且能抵禦強風。

穿過黃槿樹林，你便會到達一個小碼頭。於潮退時，在那裡的泥灘上你可以看見彈塗魚在泥面爬行及飛快彈跳。小螃蟹在橫行；而濱螺 (*Littoraria spp.*) 則在紅樹葉間棲息。沿路徑行至海灣盡處是紅樹叢與陸生植物交匯的路口，沿著山路往黃石碼頭方向繼續前行。攀上山坡後，可回望土瓜坪那片寧靜而優美的海灣。

交通：從西貢墟乘坐號巴士 (或於假日從荷李活廣場開出的96R) 前往黃石碼頭。於屋頭下車，橫過馬路後沿小徑往土瓜坪。

建議時間：約2小時

濕地考察守則

- 注意當天潮汐漲退。
- 觀察濕地生境時請保持安靜。
- 請勿捕獵、騷擾及撿取任何野生生物。
- 請勿踐踏海草及紅樹的根或幼苗。
- 請把攜來的廢物帶走。
- 尊重鄉郊傳統，愛護村民財產。
- 與野生雀鳥保持距離，不可騷擾牠們或其鳥蛋及鳥巢。

濕地考察工作紙

1) 考察地點：＿＿＿＿＿＿ 公頃

2) 列出該處找到的四種動物：

＿＿＿＿＿＿＿＿ 辨別特徵：＿＿＿＿＿＿＿＿

＿＿＿＿＿＿＿＿ 辨別特徵：＿＿＿＿＿＿＿＿

＿＿＿＿＿＿＿＿ 辨別特徵：＿＿＿＿＿＿＿＿

＿＿＿＿＿＿＿＿ 辨別特徵：＿＿＿＿＿＿＿＿

3) 列出該處找到的三種紅樹：

＿＿＿＿＿＿＿＿ 辨別特徵：＿＿＿＿＿＿＿＿

＿＿＿＿＿＿＿＿ 辨別特徵：＿＿＿＿＿＿＿＿

＿＿＿＿＿＿＿＿ 辨別特徵：＿＿＿＿＿＿＿＿

4) 紅樹在該處遇到的問題：

- ＿＿＿＿＿＿＿＿＿＿＿＿＿＿＿＿
- ＿＿＿＿＿＿＿＿＿＿＿＿＿＿＿＿

5) 紅樹怎樣適應濕地的惡劣環境?

- ＿＿＿＿＿＿＿＿＿＿＿＿＿＿＿＿
- ＿＿＿＿＿＿＿＿＿＿＿＿＿＿＿＿

6) 請建議一些保育濕地的方法：

- ＿＿＿＿＿＿＿＿＿＿＿＿＿＿＿＿
- ＿＿＿＿＿＿＿＿＿＿＿＿＿＿＿＿

3.2 Woodlands

A typical woodland

Shing Mun reservoir

Plantations along hillsides

Woodland is a common terrestrial habitat in Hong Kong. They are often dominated by trees of 3 metres in height and are characterized by a large canopy.

Nowadays, there are very few virgin forests left in Hong Kong today. During World War II, most of the trees in the primary forest were cut for fuel and wood. After the war, plantations were made in reservoir areas to prevent soil erosion. Some of examples include:

- Hong Kong – Pokfulam and Tai Tam reservoirs
- Kowloon – Kowloon reservoir
- New Territories – Plover Cove, Tai Lam, Sai Kung's High Island and Shing Mun
- Lantau Island – Shek Pik

Although planted, the woodlands share similar features of natural woodlands for studying their ecology. Apart from a few native species, planted forests mainly consist of exotic species, either in mixed species culture or in monoculture.

Plant diversity

Key Species in Woodlands

Woodland, especially in the tropics, is best known for its rich diversity of animals and plants. Different species of trees, shrubs, ferns and grasses dominate the environment. There are also many important fauna such as birds, butterflies, insects, mammals, reptiles, amphibians, spiders, etc.

Yellow flower with a bee

Vegetation

As mentioned, most forests are mixed with exotic and natural plant species. It is important to study a forest by looking at how natural it is. Basically, the more exotic trees found, the more degraded (or "unnatural") that woodland is.

Acacia confusa

As one of the most commonly planted exotic tree in Hong Kong, ***Acacia confusa*** originates from Taiwan. Growing up to 20m tall, it has long, narrow, sickle-shaped leaves with five parallel veins. Another common exotic tree is **Brisbane Box** (*Lophostemon confertus*) from Australia. This tree is at least 5m tall and has dark green leaves (15cm) with a smooth margin.

Slash Pine

Similar to the native Chinese Red Pine, **Slash Pine** (*Pinus elliotii*) is a widely planted pine for afforestation introduced from North America. With 15m tall, this tree produces small cones at the ends of the branches. In many *feng shui* woods, ***Schefflera heptaphylla*** is a common lowland tree of 4m tall. One of its distinctive features is the alternately-arranged leaves, with five to nine leaflets in a circular whorl.

Rose Myrtle

Other common plant species include:

- Shrubs – e.g. *Litsea rotundifolia*, Rose Myrtle
- Ferns – e.g. *Dicranopteris pedata*
- Herbs – e.g. orchids, grasses
- Climbers – e.g. *M. dodecandrum*, *M. micrantha*

Common Mormon

Grey Pansy

Butterflies and Insects

Hong Kong has about 250 species of butterflies. Among them, the **Papilionidae** is one of the most obvious butterfly families. These species glide with large black wings, have small hindwing "tails" and feed on nectar. Examples include Common Mormon (*P. helenus*) and Paris Peacock (*P. paris*).

The Common Tiger (*Danaus genutia*) is a bright orange butterfly with dark markings. The Grey Pansy (*Junonia atlites*) is one of many species with colorful eye-spots found on the edge of forests and in shrubland. **Skippers** are small butterflies flying at a fast direction with a characteristic perching position. Many are plain brown, but can be separated based on the spots on the underwings.

Heteropteran bug

Lantern bug

Cockroaches (*Opisthoplatia orientalis*) are common inhabitants of the forest floor. **Mantids** (*Hierodula* sp.) are most obvious with their raptorial first legs. Another crawling insect is **Black Tree Ant** (*Polyrachis dives*), which usually appears in large colony. **Grasshoppers** (*Gastrimargus marmoratus*) have long back legs, allowing them to jump around throughout the forest.

Trees with big juicy fruits often attract the small heteropteran bugs, including **Cotton Bug** (*Dysdercus cingulatus*) and **Tallow Stink-bug** (*Erthesina fullo*). Homopteran bugs are often found on the tree branches. Among them, **Lantern Bug** (*Pyrops candelaria*) is eye-catching for its peculiar elongated red "snout". **Cicadas** (*Gaeana maculata*) are famous for their loud chirping buzz and a large distinctive abdomen.

Other Animals

Woodland is considered the best habitat for **birds**. Among the 450 species recorded in Hong Kong, the common ones include Eurasian Tree Sparrow (*Passer montanus*), Blue Magpie (*Urocissa erythrorhyncha*), Spotted Dove (*Streptopelia chinensis*), Red-whiskered Bulbul (*Pycnonotus jocosus*), Chinese Bulbul (*Pycnonotus sinensis*) and Black Kite (*Milvus migrans*).

Spotted Dove

Many species of **spider** are present in Hong Kong's woodland, belonging to various families. Wood Spider (*Nephilia pilipes*) has black body with yellow spotting and a long abdomen. It can be very large, reaching greatest size of 8cm in autumn. Another hunting group is the Lycosidae, or "wolf-spiders". These spiders have a large abdomen for carrying the juveniles.

Wood Spider

As an **amphibian**, Brown Tree Frog (*P. megacephalus*) can be spotted in trees and bushes that are quite near water. As for **reptiles**, Changeable Lizard (*Calotes versicolor*) is seen on the tree trunk while the venomous Chinese Cobra (*Naja atra*) usually sneaks inside the bushes.

Lizard

Woodlands also host a number of **mammals** with abundant nutrients and good shelter. Looking carefully, one can find Masked Palm Civet (*Paguma larvata*), Chinese Pangolin (*Manis pentadactyla*), Wild Boar (*Sus scrofa*), Feral Cattle (*Bos taurus*) and Rhesus Macaque (*Macaca mulatta*).

Be aware of the fact that the animals used to human presence can become aggressive. And remember, it is illegal to feed macaques in Hong Kong.

Feral Cattle

Woodland Conservation

Damage by forest fire

With such high ecological value, woodland indeed needs to be conserved. Forest succession can be prevented by cultivation, cutting or burning. The high altitude tea terraces were abandoned, while the labour-intensive cultivation of the lower hill slopes ceased in the 1950s and 1960s. The harvesting of plants for fuel from uncultivated forests was also a major factor preventing forest succession until the 1960s, and continued in remote parts of Hong Kong until more recently.

Nonetheless, fires are the major threat to hillside wildlife. Forest fires maybe lit by people, either deliberately or accidentally, and there can be over 100 fires in a single day if the Ching Ming and Chung Yeung festivals take place in dry weather.

Hong Kong Country Parks

Covering about 45% of Hong Kong's total land, the 23 Country Parks are regarded as the "storehouse" of the territory's ecology, natural habitats, flora and fauna. As more local people and tourists visit the parks, they offer great benefits to visitors with scenery, rural heritage, recreation and nature education experiences.

Ma On Shan Country Park

Hong Kong's Country Parks vary in size, from 400 hectares (such as Aberdeen Country Park) to 2,000 hectares (such as Ma On Shan Country Park) to over 4,000 hectares (such as Sai Kung East Country Park). These protected areas are essential for Hong Kong's sustainable development by allowing wise use of land resource for present and future generations.

Aberdeen Country Park

Various conservation methods, such as seed collection, cutting and air layering, have also been applied to propagate rare and endangered plants. The seedlings are produced with care in nurseries for replanting in their potential natural habitats in various country parks. Successful examples of propagation include *Keteleeia fortunei*, *Camellia crapnelliana* and *Camellia granthamiana*.

To assist in maintaining habitat diversity and species richness, tree-planting schemes have been designed to enhance biodiversity rather than purely increase erosion control. Mixed planting, rather than monoculture, has been adopted. Moreover, using selective plantings to help maintain diverse habitats, native trees are planted to enrich ecological areas.

For more details, please visit AFCD's website (www.afcd.gov.hk/parks/parks.htm).

Suggested Woodland Field Sites in Hong Kong

Woodlands can be easily found in Hong Kong countryside, especially in the catchment areas of main reservoirs, such as Pokfulam, Tai Tam, Tai Lam Chung, Shing Mun and Shek Pik. Woodlands in **Tai Po Kau Nature Reserve**, **Sha Lo Tung** and **Lung Mun Country Trail** are comparatively mature ones worth to study in Hong Kong.

Entrance of nature reserve

Ecological richness

Stick Insect

Brown Tree Frog

Trails in the woodland

1) Tai Po Kau Nature Reserve

As Hong Kong's most well-preserved woodland, Tai Po Kau Nature Reserve is situated at the eastern side of Tai Mo Shan Country Park near Tai Po Market Station. Within the area of 460 hectares, there are four coloured trails and one nature trail (the shortest), running along the valleys of various heights and lengths.

Tai Po Kau has been designated as a **Nature Reserve** rather than a Country Park because of its conservation interests in wildlife biodiversity, natural habitat and its large size. The well-protected forest offers different ecological niches for different forms of wildlife. These habitats include an open orchard area, freshwater habitats in various streams, dense canopies, as well as the humid and shaded forest floors.

The main planted tree species is Chinese Red Pine, giving rise to the area's name of Tsung Tsai Yuen ("Pine Garden"). Besides, Camphor Tree, China Fir, Acacia and Paper-bark Tree have also been planted.

In this woodland heaven, over 160 bird species have been recorded. Magpie, Chinese Bulbul and Crested Bulbul are commonly found. Tai Po Kau is the home of Wagtails in winter, while in summer Cuckoos and Black Drongos are plentiful. Visitors are recommended to use binoculars for viewing birds and move quietly along the forest paths.

Tai Po Kau hosts other wildlife including:

- Dragonflies – *Ceriagrion auranticum*
- Butterflies – Common Birdwing, Spotted Sawtooth
- Insects – Blue Spotted Tiger Beetle and Stick Insect
- Amphibians and Reptiles – Brown Tree Frog
- Mammals – Leopard Cat

Transportation: KMB buses no. 70, 72, 72A, 73A and 74A all pass through Tai Po Road (drop off point at Chung Tze Yuen).

Suggested Time: 3-5 hours at the site

2) Sha Lo Tung

Sha Lo Tung occupies more than 80 hectares of land area, and the basin is surrounded by Pat Sin Leng Country Park. About 20 years ago, some property developers purchased land from local villagers, and intended to turn these farmlands and villages into golf courses and low-density residential areas. Yet due to fierce oppositions from green groups and lack of support from the government, Sha Lo Tung has remained rural.

Entrance of Sha Lo Tung

Because of its abundant ecological resources and cultural value, Sha Lo Tung has been widely recognized by scientists and green groups. The area has been established as a "**Site of Specific Scientific Interest**" (SSSI), and any local development is strictly limited and highly regulated. The valley at Sha Lo Tung has managed to sustain many unpolluted streams and dense *feng shui* forests, which provide an ideal environment to nurture a wide variety of plants and animal species.

Feng Shui forests

There are about 170 herb species and 600 vascular plant species in this small area. More than 100 bird species have been spotted here. Large mammals such as wild boar, barking deer and pangolin were found. The site also has **tremendous dragonfly and amphibian species**.

Woodlands

Another focus of Sha Lo Tung is its cultural value. The old villages provide hints to the living culture of Hong Kong's early settlers. Although most village houses are now empty and deteriorating, there are traces of traditional lifestyle of local villagers years ago. The broken floor, beds, farming and cooking tools left are concrete evidences of history. In fact, the Hong Kong Government has labeled the abandoned "Hakka" village houses as second grade historical relics.

Hakka village

On 12 January 2001, the Town Planning Board announced a new policy for Sha Lo Tung, protecting over 70% of the natural landscape, raising nature conservation and preserving the rural characteristics.

Hiking trails heading towards Tai Po

Transportation: There is no direct transportation to Sha Lo Tung. It is advised to take a taxi from Tai Po Market Station and follow along Fung Yuen Road. There are several hiking trails in the area.

Suggested Time: 3 hours at the site.

3) Lung Mun Country Trail

Tai Mo Shan

Cicada

Macaque

Trail ends at Shing Mun

Located in central New Territories, Lung Mun Country Trail is a 4-km walk in between Chuen Lung of Tai Mo Shan Country Park and the Pineapple Dam of Shing Mun Reservoir.

At 957 metres elevation, Tai Mo Shan ("Big Hat Mountain" in Chinese) is the highest peak in Hong Kong. Limited by climatic and geographic factors, the plantations end at the 550-metre contour, above which shrubs and grasses dominate the landscape.

Most of the trees in the area were cut down during the Japanese occupation in World War II. Extensive reforestation was carried out after the war, mainly using Chinese Red Pine. Slash Pine, Brisbane Box, Cajuput-tree and Acacia were planted later.

Since the park is close to the densely populated urban area of Tsuen Wan, wildlife is not abundant. Nonetheless some common birds, squirrels, wild boar and pangolins are seen occasionally. The protected **Hong Kong Newt** may be found in the park's streams.

The long-tailed Macaque inhabiting the area are not native wildlife but descendants of monkeys released in 1920. They eat leaves, fruits, insects and other small animals. **Feeding the monkeys is strictly prohibited** and is also dangerous, as they can inflict painful bites.

Transportation: Take Bus No. 51 at Tsuen Wan MTR station towards Kam Tin. Get off at Chuen Lung, cross the road and follow the signs along the road path inside the village. After a 15-minute walk, you'll reach the beginning of the trail.

Suggested Time: 3 hours at the site

General Guidelines for Woodland Field Trips

Exploring inside woodland

- Keep away from cliffs and steep slopes.
- Always check the weather beforehand. Avoid visiting woodlands when it rains because the paths or slopes may be very slippery.
- Wear hiking boots and long pants.
- Do not smoke or start a fire at any time.
- Use sun block and mosquito repellents for protection.
- Do not hunt, disturb or collect any wildlife.
- Take plenty of water and drink frequently to prevent dehydration.
- Respect village traditions and treasure villagers' properties.

Practical Worksheet for Woodland Field Trips

1) Area of site visited: ______________ hectares

2) Name 4 animal species that you found in the site:

___________________ Adaptive features: __________________________

___________________ Adaptive features: __________________________

___________________ Adaptive features: __________________________

___________________ Adaptive features: __________________________

3) What are the different phases of polymorphism of insects? Briefly explain.

____

____

4) What is forest succession? Explain briefly.

____

____

5) Explain the importance of soil and leaf litters in woodland.

____

____

6) Comment on the effects of human activities on the woodland ecosystem.

____

____

3.2 林地

林地是香港常見的陸地生境之一，雖然大部分為植林，但我們仍可藉以學習考察一般林地生態的方法。

樹林一般都是由一些高於三米，並有茂密樹冠層的樹木組成，而典型的灌木林植物如崗棯 (*Rhodomyrtus*) 大都絕迹於樹林，相反，木質攀緣植物在原生樹林則甚為常見。然而，香港已沒有原生樹林保留至今。

第二次世界大戰期間，原生林的樹木被大量砍伐，作燃料及木材之用。植林的工作在戰後展開，主要集中於新興建的水塘區域，以緩減土壤侵蝕，其中包括：

- 香港　- 薄扶林水塘和大潭水塘
- 九龍　- 九龍水塘
- 新界　- 船灣淡水湖、大欖涌水塘、西貢萬宜水庫及城門水塘
- 大嶼山 - 石壁水塘

隨著香港經濟起飛，民生改善，農業日漸息微，香港部分土地的生態價值得以提高，但另一方面，廣泛的發展需求正威脅著某些地區的生態，致使香港的環境得失參半。

林地的主要物種

林地以擁有豐富而多樣化的動物和植物物種見稱，遷徙鳥和紅樹林無疑都具有重要的生態價值，此外，這裏亦孕育著其他重要的動植物如水生植物、哺乳類、爬行類、兩棲類、淡水魚類、蝴蝶、蛾及蜻蜓等等。

植物

植林大多採用外來樹種，可分為只選用單一品種的純林或不同品種的混交林。然而，外來樹種有時亦可在天然林內找到，而植林內亦可發現本土樹種。一個樹林是否天然而成對研究來說是很重要的。樹林種植了越多外來品種，其生態價值便越低(非自然)。

台灣相思 (***Acacia confuse)*** 的原產地是台灣，是香港最廣泛栽種的植林樹種之一，可高達20米，葉片呈鐮刀狀披針形，有5條平行葉脈。另一外來品種是原產澳洲的**紅膠木** (*Lophostemon confertus*)，高5至15米，葉片呈深綠色，全緣，長15厘米。

原生的馬尾松及由南美引入的濕地松都是本地常見的植林品種，後者可高達15米，細小的球果生於枝條末端。而鴨腳木 (***Schefflera heptaphylla***) 則是風水林常見的低地樹種，高3至4米，掌狀複葉，互生，有5至9枚小葉。

蝴蝶及其他昆蟲

本港大約有240種蝴蝶，鳳蝶 (Papilonidae) 是最常見的品種之一，其翅膀以黑色為主，後翅有突尾，喜歡吸食花蜜。例子有玉斑鳳蝶 (*P. helenus*) 和巴黎翠鳳蝶 (*P. paris*)。

虎斑蝶 (*Danaus genutia*) 雙翅以鮮橙色為主，並有黑色翅脈。美眼蛺蝶 (*Junonia almana*) 是其中一種翅膀上具有眼斑的蝴蝶，常見於樹林邊緣及權木叢。弄蝶體型細小，飛行迅速，喜歡以直線飛行。弄蝶的翅膀以褐色為主，但可從其翅底的斑紋來分辨牠們。

蟑螂 (*Opisthoplatia orientalis*) 的體型小於3厘米，是森林地面常見的物種，螳螂 (*Hierodula* sp.) 有捕食用的前腿。其他爬行昆蟲有雙齒多刺蟻 (*Polyrachis dives*)，牠們常以大型的群落出現。草蜢 (*Gastrimargus marmoratus*) 擁有碩長的黑色後腿，可讓牠們在森林中跳躍。

許多長有豐富多汁的醬果樹木都會吸引一些細小的異翅目昆蟲，如棉紅蝽(*Dysdercus cingulatus*) 和麻皮蝽 (*Erthesina fullo*)。同翅目昆蟲則常見於樹枝上，其中尤以龍眼雞 (*Pyrops candelaria*) 奇特的伸長紅色「喙」最為引人注意。蟬以其響亮而嘈雜的鳴聲最為人熟悉，牠們還擁有明顯的腹部。

鳥類及其他動物

林地被認為是鳥類最理想的生境，香港現時記錄到450種鳥類，其中最常見的包括麻雀(*Passer montanus*)、紅嘴藍鵲(*Urocissa erythrorhyncha*) 、珠頸斑鳩(*Streptopelia chinensis*)，紅耳鵯(*Pycnonotus jocosus*)、白頭翁(*Pycnonotus sinensis*)和黑鳶(*Milvus migrans*).

香港林地可找到許多分屬不同科的蜘蛛。斑絡新婦 (*Nephilia pilipes*) 是黑色，帶有黃色斑點，有長形腹部，秋天時長到最大，體型可以十分巨大。另一獵食群種是狼蛛 (Lycosidae)，這種活躍的蜘蛛會把幼體放在肥大的腹部。

斑腿泛樹蛙(*Polypedates megacephalus*)是兩棲類的一種，常見於附近有水源的樹木和灌木間。爬蟲類如變色樹蜥 (*Calotes versicolor*) 常見於樹幹上，而有毒的眼鏡蛇(*Naja atra*) 則常藏身於灌木叢中。

林地還為各種哺乳類動物提供大量養料和棲息地。若我們仔細留意，可發現果子狸 (*Paguma larvata*)、穿山甲 (*Manis pentadactyla*)、野豬 (*Sus scrofa*)、黃牛 (*Bos taurus*)和獮猴 (*Macaca mulatta*)。

遊人要謹記不怕人的動物可以變得很具侵略性，而餵飼獮猴是會觸犯香港法例的。

林地保育

耕作、砍伐和山火都會阻礙森林的演替。海拔較高的茶樹梯田已經廢棄了超過150年，而在較低山坡上的勞工密集式農業操作也在1950年代和1960年代終止。1960年代以前，人們在沒有開墾的山坡上搜集植物以作燃料，這亦是森林演替受阻的一個重要因素，直至現時為止，香港一些偏遠地區仍然會出現這種情況。現在，山火是山坡上的植被和野生動物所面對的最嚴重威脅。無論是蓄意或者無意，香港所有山火都是人為造成的。假如清明節和重陽節正好碰上乾燥的天氣，單單在一天裏，就可以發生超過一百場山火。

香港郊野公園

本港共有23個郊野公園，佔全港總面積百分之四十五，是保存本地生態、自然生境及動植物的重要範圍。近年到訪郊野公園的市民及旅客日漸增加，他們可欣賞郊野怡人的風景及自然資產，並進行康樂活動，更可獲得有關自然教育的體驗。

香港郊野公園範圍大小不一，面積較小的只有400公頃 (如香港仔郊野公園)，較大的佔地2,000公頃 (如馬鞍山郊野公園)，甚至超過4,000公頃 (其中最大面積之一的西貢東郊野公園)。這些受保護地區規劃了現代和後代對土地資源的運用，對香港的可持續發展十分重要。(網址: www.afcd.gov.hk/parks/parks.htm)

當局亦以不同的方法如採集種子，插枝及高空壓條法繁殖稀有及瀕危植物，並將種子在苗圃發芽育苗，再將樹苗種植於郊野公園內適宜的自然生境。油杉、克氏茶及葛量洪茶都是人工繁殖的成功例子。為了維持生境多樣性和物種多樣化，植樹計劃現以增加生物多樣性為目標，而非單純為了控制土壤侵蝕，因而採用了混合樹種，而非單一種植。

香港林地考察地點

我們可在香港的郊野公園輕易找到林地，特別是主要水塘的集水區如薄扶林、大潭、大欖涌、城門和石壁。大埔滘自然保育區、沙羅洞和龍門郊遊徑都是香港較成熟的林地，但這三片林地均位於山坡，遊人前往時必需小心。

1) 大埔滘自然保育區

大埔滘自然保育區位於大帽山郊野公園的東面，鄰近大埔墟火車站，是香港現存保留得最好的林地。該地區面積達460公頃，共有四條用顏色指引的小徑及一條自然教育徑，每條小徑長度和高度不一，沿山丘婉蜒伸展。

由於大埔滘面積廣，且具有多樣性的生物和自然生境，需要加以保育，因此被劃定為自然保育區而非郊野公園。這裡的森林保存良好，為各式各樣的生物提供不同的生境，如保育區邊緣的果園、溪澗的淡水生境、茂密的樹冠層，以及遮蔽的森林地面。

由於這裡種植了大量馬尾松，因此又被稱為松仔園，而其餘栽種的樹種包括樟、杉、台灣相思及白千層。

這兒亦是林鳥天堂，記錄了超過160種鳥類，常見的品種包括喜鵲、白頭翁和高雞冠。鶲鴒於冬天棲息在這裡，而夏天則有大量的杜鵑和黑捲尾。遊人宜配備雙筒望遠鏡，沿森林內的小徑靜心觀鳥。

大埔滘是許多野生生物的棲息地，如：蜻蜓 (琉球橘黃蟌)、蝴蝶 (裳鳳蝶和鋸粉蝶)、昆蟲 (藍點甲蟲)、兩棲類及爬蟲類 (斑腿泛樹蛙) 及哺乳類 (豹貓)。

交通：公共巴士路線70, 72, 72A, 73A 及74A 都會途經大埔公路 (於松仔園下車)。

建議時間：約3–5小時

2) 沙羅洞

沙羅洞的面積超過80公頃，其盆地被八仙嶺郊野公園環抱。發展商於二十年前向村民買地，打算將農地改建為哥爾夫球場及低密度住宅，但由於此計劃遭到綠色團體的強烈反對及缺乏政府支持，沙羅洞現時仍保留著郊區風貌。

沙羅洞具有大量的生態資源及文化價值，被科學家和綠色團體高度評價，其中大部分範圍已列為「具特殊科學價值地點」(SSSI)，當地的發展受到嚴格限制和高度規管。沙羅洞有多條未受污染的山溪及茂密的風水林，為孕育各類不同的動植物提供理想的生境。

在這小小的地方，草本植物多達170種，維管束植物則有600種。這裡錄得超過100種鳥類，亦可找到一些大型哺乳類動物的踪迹，如豪豬、赤麂和穿山甲。這區亦有不少蝴蝶和兩棲類。

此外，沙羅洞具有很高的文化價值，這裡古老的村落見証著香港早期定居者的生活文化。雖然大多村屋現已荒廢及失修，但當你一踏足此地，便會感受到村民昔日的生活。現已破爛的地板、床、農具和煮食用具，都是歷史的見證。事實上，香港政府已將這裏荒廢的客家村屋列為二級古蹟。

城市規劃委員會於2001年1月12日就沙羅洞宣佈新政策，以保護這裡超過百分之七十的自然景觀，加強自然保育及保留鄉郊特色。

交通：現時並沒有交通直達沙羅洞，遊人可從大埔公業邨乘搭的士前往，該處有數條行山徑。

建議時間：約3小時

3) 龍門郊遊徑

龍門郊遊徑位於新界，由大帽山郊野公園的川龍起步，步行往城門水塘的菠蘿霸，全程約四公里。

大帽山高957公尺，是全港最高之山峰，基於氣候和地理環境影響，林區衹能伸展至五佰五十公尺高度，再高的地方則以灌木叢及草叢為主。

於二次大戰日軍侵佔香港期間，這裡的樹木被大量砍伐，植林的工作於戰後廣泛展開，早期主要採用馬尾松，後期則引入濕地松、紅膠木、白千層及台灣相思。

由於該區鄰近人口稠密的荃灣，因此野生生物並不算多，但常見的鳥類、松鼠、豪豬和穿山甲亦偶有發現。而受法例保護的香港瘰螈亦可在這裡的溪流找到。

這裏的長尾獼猴並非本土野生動物，而是一九二零年代在此處放生的獼猴後代，牠們主要進食葉、果、昆蟲和其他小型動物。由於這些獼猴時會向人襲擊，因此餵飼牠們是十分危險的，而且亦會觸犯法例。

交通：於荃灣地鐵站乘搭51號巴士往錦田，於川龍下車，橫過馬路後，按路徑及指示入村，步行約15分鐘後，便到達郊遊徑的起點。

建議時間：約三小時

林地考察守則

- 考察前應查明天氣狀況，避免下雨時出發，因為路徑會變得十分濕滑。
- 穿著登山鞋和長褲。
- 戴帽子和太陽眼鏡，但進行觀察時應脫下眼鏡。
- 切勿狩獵、騷擾或取走野生生物。
- 帶備充足食水並經常飲用，防止脫水。
- 帶備驅蚊用品和急救箱。
- 塗抹防曬用品以防曬傷。
- 切勿吸煙和生火。
- 遠離懸崖和斜坡。
- 尊重鄉村傳統，珍惜村民財物。
- 與野生雀鳥保持距離，不可騷擾牠們或其鳥蛋及鳥巢。

林地考察工作紙

1) 考察地點： ________________ 公頃

2) 列出四種於該處找到的動物：

__________________________ 適應特徵： ____________________________________

__________________________ 適應特徵： ____________________________________

__________________________ 適應特徵： ____________________________________

__________________________ 適應特徵： ____________________________________

3) 昆蟲有哪種變態呢? 試簡單說明。

____

4) 試解釋泥土與枯枝落葉對樹林的重要性。

____

5) 評論人為因素對於樹林生態系統的影響。

____

3.3 Streams and Shorelines

Hong Kong has around 1,650 square kilometres of sea area including a wide variety of different marine habitats which contain a great diversity of marine life. Its coastline is around 800 kilometres long and supports a great variety of coastal and marine ecosystems. These include rocky shores, sand-flats (sandy shores) and mudflats (muddy shores). Mangrove communities and sea-grass beds can be found on the intertidal shores. The distribution and the great diversity of water ecosystems in Hong Kong are related to several different ocean currents and to freshwater water discharges from the Pearl River Delta into the western Hong Kong waters.

There are different tidal zones along the shores: the **Splash Zone**, the **Intertidal Zone** and the **Subtidal Zone**. The Intertidal Zone can be divided into high, mid and low intertidal depending on the relative times that these areas spend underwater: the high intertidal zone is exposed to air for long periods twice per day, the mid-intertidal zone is exposed to air briefly once or twice per day, and the low intertidal zone is exposed only during the lowest tides.

Beautiful shorelines

The splash zone above the high tide line is almost always exposed to air, and receives most of its water in the form of spray. The subtidal zone is the coastal marine area below the intertidal zone, that is, below the lowest tide line of the year. Both the tidal cycle and the seasonal variation affect the distribution of the intertidal species.

Streams

There are no large rivers in Hong Kong due to the scarcity of flatlands caused by rising sea levels following the last glaciation around 10,000 years ago.

However, small rivers, hill streams, lowland streams and even some artificial channels support a rich biota, including a great variety of fish, benthic invertebrates, and mammals adapted to riverine wetlands.

Hong Kong's stream organisms are greatly affected by the seasonal Monsoon rainfall. The flow increase caused by these rains can reduce animal populations through drift as well as reducing algal densities through abrasion; also, there is a strong relationship between benthic animals and the sediments that they live on, and water currents determine the particle structure of streambeds so therefore they also determine the distribution and abundance of the organisms that inhabit streams.

Streambed particles of various sizes at different positions of a stream channel are related to different water current forces: In shallow areas with a relatively fast water current, such as riffles, the stream beds are mainly composed of boulders and cobbles whereas in areas with slower or reduced water currents, such as pools, they are mainly composed of fine particles such as silt, clay and fine sand.

Key Species in Shorelines

There are over 200 islands in Hong Kong waters and various different marine or coastal organisms live along their various types of shores; local waters contain a very wide diversity of species of fish, crustaceans, mollusks and other marine life. This diversity is greatly due to the fact that Hong Kong is in the sub-tropical region and provides a habitat for species belonging to both tropical and temperate climates, as well as to the fact that the western Hong Kong waters receive freshwater directly discharged from the Pearl River whereas the eastern waters are hardly influenced by the river outflow and have predominantly oceanic characteristics.

Vegetation
Many plants, which have adapted to the different coastal conditions, can be found along the Hong Kong coastline:
Sand-binding plants such as *Vitex rotundifolia, Spinifex littoreus* and *Wedelia prostrate*, as well as various bushes such as *Clerodendrum inerme, Pandanus tectorius, and Scaevola sericea can be found o*n the seaward side of the sandy shores, and shrubs or trees of the *Hibiscus tiliaceus, Macaranga tanarius,* and *Cerbera manghas* species can be found *on the* landward side.

Cerbera

Algal species

Creepers such as *Asparagus cochinchinensis, Morinda parvifolia, Canavalia maritima, Ipomoea pes-carprae* and shrubs such as *Scaevola sericea, Lysimachia mauritiana, Vitex rotundifolia, Pandanus tectorius* and *Phoenix hanceana* can be commonly found behind rocky shores.

Lichens dominate the environment in the splash zone, and cyanobacterium *Kyrtuthrix maculans* acts as a boundary to separate the high and mid intertidal zones. The mid intertidal zone is dominated by encrusting algae such as *Ralfsia expansa* whereas erect algae, such as *Sargassum hemiphyllum,* are commonly found in the subtidal zone.

Kandelia obovata

Dwarf mangroves such as *Kandelia obovata, Aegiceras corniculatum, Avicennia marina and Acanthus ilicifolius* are quite common along muddy shores.

Fauna

Hard coral and reef-associated fishes

By international standards, Hong Kong is considered to have an extremely rich coral diversity, predominantly in the eastern waters, consisting of some 80 species of stony coral and at least 14 species of soft coral which can be found on the boulders and bedrocks in shallow waters along its rocky shores. What we see as branches are actually whole colonies of polyps, which are small animals that belong to the *Phylum Cnidaria*, with a hard skeleton and a soft body. They are usually attached to rock, or to the skeletons of dead polyps. The flower-like formations that can be seen on the coral are actually the tentacles with which they capture food from the water.

The hardiest of the intertidal organisms, small creatures like grazing littorinid snails and sea-laters (*Ligia exotica*), can be found in the splash zones, and grazing limpets (*Collisella dorsuosa*) and filter-feeding barnacles (*Chthamalus malayensis*) can be found colonizing more exposed parts along the rocky shores. When the tide is high, the limpets move about grazing on algae. As the tide drops, they usually return to their resting place.

Sea turtles

Chitons (*Acanthopleura japonica*) are highly desiccation-tolerant and thus well-suited for life on the rocky shores, being efficient grazers, and capable of excavating both encrusting algae and microalgal biofilm for food. They, as well as small barnacles (*Tetraclita japonica*, *T. squamosa*) and large barnacles (*Megabalanus volcanoare*) are quite abundant along different regions of the intertidal zones.

Snails and bivalves

Mussels and sea urchins can be found along semi-exposed shores. Sea urchins living among mussels are protected from predators and require more wave force to dislodge them.

Chiton

A variety of species of marine fish also breed in Hong Kong waters. The red sea bream (*Chrysophs* major), one of the several sea bream varieties whose fry are abundant along the shore of Mirs Bay in early spring, is a typical example of marine life in the eastern waters, and one of the most common fish around exposed rocky shores in Hong Kong is the stellar rockskipper, *Entomacrodus stellifer lighti.*

Key Species in Streams

Hong Kong Newt

Rivers and streams are flowing-water ecosystems which are characterized by the unidirectional flow of water from their sources (where they begin) to their mouth (where they enter into another water body). They cannot be isolated from their terrestrial surroundings, and since the types of land where water flows differ widely, rivers and streams always undergo great changes between their sources and mouths.

Aquatic plants, insects, snails, crustaceans, freshwater fish, amphibians and reptiles are commonly found in Hong Kong streams. Some examples are *Marcopodus hongkongensis* and the Hong Kong Newt (*Paramesotriton hongkongensis*).

Ficus spp.

Dragonfly

Damselfly

Macrobrachium hainanense

Gambusa affinis

Vegetation

Stream plants are freshwater plants that are found either at the margins of water bodies, or as emergent forms in shallow waters and submerged or floating forms in deeper waters: Those found at water margins are shrubs and trees which are not truly aquatic but that need moisture or a continuous water supply to live, such as *Ficus fistulosa*, *Ficus hispida* and *Ficus pyriformis*.

The emergent, submerged and floating leaved stream plants are plants that have adapted to the running water conditions and can be found in clumps in hill-streams, like Filamentous algae, mosses and liverworts.

Fauna

Most of the fauna living along Hong Kong streams are insects, though crustaceans, mollusks, amphibians and small fish can be commonly found also.

All stream insects are amphibiotic, that is, they spend their larval stage in water and their adult stage on land. Due to the stable water temperatures, many of these insects, such as mayflies, stoneflies (Plecoptera) such as *Kamimuria*, and caddisflies (Trichoptera) such as *Ganoema extensum*, as well as many dragonflies and damselflies (Odonata) can breed all year round. The damselfly (*Euphaea decorate* and *Rhinocypha perforata*) and the largest, most common water skater (*Ptilomera tigrina*) can be easily found along Hong Kong hill-streams.

Other small creatures, such as snails (*Brotia hainanensis*), bivalves or fresh water clams (*Corbicula fluminea*) and spiders such as the Big-Jawed Spider (*Tetragnathidae mandibulata*) with its slender body and long thin legs to move along the boulders, are quite abundant and widespread along Hong Kong streams.

The Freshwater crab (*Cryptopotamon anacoluthon*) and the Mitten crab (*Eriocheir japonicus*) live under the rocks along the stream banks; the Hong Kong Newt (*P. hongkongensis*) and the endangered Romer's frog can be found on the streambeds if you look carefully.

Small fish, such as *Parazacco spilurus, Pseudogastromyzon myers*, and *Ctenogobius duospilus* are also abundant in Hong Kong. The most common and most widespread shrimp species in Hong Kong freshwaters, the atyid shrimp (*C. cantonensis*) and *M. hainanense* can be commonly found thriving on dead leaves in pools.

Marine Environment Conservation

Rocky shore

Hong Kong's economic development and population growth has limited the water mass by channelling or damming it for human use, thereby threatening the survival of acuatic life. Much of the original flora and fauna has been eliminated because of this. The marine parks and marine reserves programme is a conservation programme intended to protect the existing marine environment.

The **Marine Parks Ordinance (Caps. 476)** was enacted in 1995 for Hong Kong's marine conservation, and provides for the designation, control and management of marine parks and marine reserves. The Marine Parks and Marine Reserves Regulation was enacted in July 1996 to provide for prohibition and to control certain activities in marine parks or marine reserves.

A marine park is a relatively large area in Hong Kong waters that is intended for conservation purposes only, though recreation activities may be carried out as long as they are non-destructive to the marine environment. A marine reserve is a smaller area but with a high conservation value, thus most activities ar e prohibited except for scientific and educational study.

The following activities are prohibited within marine parks and marine reserves:

- Unauthorized fishing, hunting, collecting or possession of any wild animals or plants, or their parts.
- Disturbing the eggs, juveniles or the nesting sites of any protected marine species.
- Possession of a trawl net, spear gun, explosives, chemicals or any devices using electric charge which may be used for fishing.
- Water skiing, jet skiing or water scootering.
- Operating vessels exceeding the speed of 10 knots.
- Damaging any shoreline features on a beach, mudflat, cliff or seabed.
- Damaging notices, markers or other installations used by the Authority.
- Obstructing waterways, polluting water bodies or littering.
- Unauthorized commercial activities, group activities, mooring and anchoring.
- Conducting any mariculture operation.

Swimming, diving and boating are also prohibited in marine reserves.

Boulders and cobbles

Protect the shorelines

Visiting by boats

Suggested Field Sites in Hong Kong

At present, there are four marine parks and one marine reserve in Hong Kong:

- The Hoi Ha Wan Marine Park, the Yan Chau Tong Marine Park and the Cape D'Aguilar Marine Reserve designated as such on 5 July 1996.
- The Sha Chau and Lung Kwu Chau Marine Park designated as such on 22 November 1996.
- The Tung Ping Chau Marine Park designated as such on 16 November 2001.

In this chapter, we recommend you to visit **Tung Ping Chau Marine Park**, **Hoi Ha Wan Marine Park** and **Tai Ho Stream**.

Rocky shore

Sandy shore

Sandwich-like stone

Bivalves, snails and algae at intertidal area

1) Tung Ping Chau Marine Park

The Tung Ping Chau Marine Park is the fourth Marine Park in Hong Kong and was designated as such on the 16th of November 2001. Its sea area covers 270 hectares enclosing Ping Chau island in the Mirs Bay in the northeastern waters of Hong Kong.

Coral communities, algae beds and rocky shores are important marine habitats found in Tung Ping Chau. Over 45 species of hard corals have been recorded in the inner Tai Tong Wan and A Ma Wan bays, as well as over 130 species of reef-associated fish from 41 families such as butterfly fish, parrotfish, groupers, snappers, sweetlips and filefish and over 100 species of marine invertebrates. Two "Core Areas", which are areas that receive stronger protection in order to conserve important biodiversity, are found within this marine park: Tai Tong and A Ma Wan.

More than 40 species of marine algae can also be found in Tung Ping Chau Marine Park. Along the coast of Lung Lok Shui, you can find the famous Brown Algae (*Sargassum* sp.), Red Algae (*Hypnea* sp.) and Green Algae (*Ulva* sp.) which are considered as the best algal beds in Hong Kong.

Tung Ping Chau Park's rocky shores provide opportunities for rock formation and intertidal rocky shore ecological studies. You can find lichens, encrusting algae and littorine snails on the upper shores, while Acon barnacles (*Tetraclita japonica*), limpets, top shells, chitons and mussels (*Perna viridis*) can be found in the intertidal zones.

Transportation: The ferry can be taken from Ma Liu Shui Public Pier (located near the University KCR station) to Tung Ping Chau Pier, on Saturdays, Sundays and public holidays.

Suggested Time: 6 hours at the site

2) Hoi Ha Wan Marine Park

Hoi Ha Wan Marine Park is a sheltered bay with pristine water quality that provides a good marine environment for a great variety of marine organisms, which is located north of the Sai Kung West Country Park. The seaward boundary of the park is an imaginary line drawn from the tip of Heung Lo Kok to the tip of Kwun Tsoi Kok over the northern end of Flat Island (Ngan Chau) and the northern end of Moon Island (Mo Chau).

Hoi Ha peninsula

It covers about 260 hectares of sea area and boasts well populated coral communities including 59 of the 88 stony coral species recorded in Hong Kong. Hard corals of the *Cyphastrea spp.*, *Echinophyllia spp.*, *Favia spp.*, *Favites spp.*, *Goniopora spp.*, *Leptastrea spp.*, *Lithophyllon spp.*, *Pavona spp.* and *Platygyra spp.* species are predominant in either one or more coral sites within the Marine Park.

Seashore at Hoi Ha

Over 120 species of reef-associated fish such as groupers, snappers and parrotfish have been recorded in this Marine Park, besides a wide variety of marine animals such as sponges, tube-worms, soft corals, sea cucumber crabs, starfish and jellyfish.

Mangroves along the shore

Various true mangrove species including *Kandelia candel*, *Aegicera corniculatum*, *Excoecaria agallocha* and *Bruguiera gymnorrhiza*, *Avicennia marina* and *Lumnitzera racemosa* have been recorded in Hoi Ha Wan Marine Park. Mangrove areas commonly act as nursery grounds for juvenile fish and other intertidal and subtidal invertebrates.

Grapsus albolineatus

Seahorse

Transportation: Take **Green minibus No. 7 from** Sai Kung Town to Hoi Ha Village or take **KMB No. 94 from** Sai Kung Pier to Wong Shek Pier and get off at Ko Tong and **walk along the Hoi Ha Road** for about 60 minutes.

Suggested Time: 6 hours at the site.

3) Tai Ho Stream

Tai Ho Stream

Gulf of Tai Ho

White Wagtail

Moth

Tai Ho is located on the northeastern part of Lantau Island. It includes a variety of different types of ecological habitats such as *feng shui* woods, shrubland, grassland, streams and seasonal marshes as well as a bat roosting site in the upland areas, and mangrove, sea grass beds and mudflat habitats in the lowland estuary.

Tai Ho Stream is the home of the highest number of freshwater fish species in Hong Kong, fifty three species, causing it to be listed as a "Site of Special Scientific Interest" on the 5th of May, 1999. Also, 10 amphibian species live in the stream, including the endemic Romer's Tree Frog, *Philautus romeri*. Horseshoe Crabs, which are now relatively rare, used to inhabit the western shore of Tai Ho, where the rare sea grass *Halophila beccarii* still grows.

Many flying species, such as Little Egret, Crested Myna, Black-capped Kingfisher and White Wagtail are commonly found in the cove of Tai Ho, and since there are many flowering plants growing along both sides of the stream, moths and butterflies are quite common in the area.

Transportation: Take New Lantau Bus No. 36 (circular route) from Tung Chung Town Centre to Siu Ho Wan Vehicle Detention Pound. Ride for 5 minutes and get off at Cheung Tung Road. The timetable for this route from Tung Chung Town Centre is: 7:45am, 10:00am, 3:00pm, 5:00pm and 7:30pm daily.

Suggested Time: 3 hours at the site

General Guidelines for Stream/shoreline Field Trips

- Do not overturn stones and boulders.
- Do not hunt or disturb any marine wild life.
- Do not litter.
- Do not pollute waters.
- Respect our marine or freshwater environment.

Conducting field trip in streams

Practical Worksheet for Stream/shoreline Field Trips

1) Area of site visited: ______________ hectares

2) Name 3 animal species that you found in the site:

____________________ Distinctive features: _________________________

____________________ Distinctive features: _________________________

____________________ Distinctive features: _________________________

3) Name 3 plants that you found in the site:

____________________ Distinctive features: _________________________

____________________ Distinctive features: _________________________

____________________ Distinctive features: _________________________

4) Problems encountered at shorelines / streams in the site:

- ____
- ____
- ____

5) How can animals survive in the tough conditions of intertidal zones/streams?

- ____
- ____

6) Please suggest some of the ways to conserve shorelines / streams:

- ____

3.3 河溪和海岸

香港海域面積佔1,650平方公里，擁有一個不同的海洋生態環境，使海洋生物多樣性更為豐富。其海岸線長達800公里，使香港沿岸及海洋的生態系統豐富起來，其中包括硬質海岸如岩岸及軟泥灘如沙灘和泥灘，而紅樹林和海草床亦可在潮間帶輕易找到。而本港海洋生態系統的分佈和多樣性受不同的海洋水流及來自珠江三角洲流入香港西面水域的淡水影響。

海岸可劃分**濺水帶**、**潮澗帶**和**潮下帶**等不同的潮汐區，潮澗帶可依據被水淹沒的時間長短分為高、中、低潮澗帶。高潮澗帶每天有兩次長時間暴露於水面；中潮澗帶只有一兩次現於水面多數在中度退潮時；而低潮澗帶只有在潮水退到最低時才暴露出來。在濺水帶高於最高潮漲線大多數時間都暴露於水面，給浪花水濺。而潮下帶是位於潮澗帶以下岩岸海洋範圍，一般都在全年最低潮退潮線以下。汐漲退周期及季節更替影響沿岸潮澗帶物種的分佈。

由於一萬年前冰河時期後水平線上升，引致缺乏平原，因此香港沒有大河流，而上述的生態系統主要由小河流、山澗、低地河溪以及人造河道來維持。這些河溪生態系統孕育了豐富的動植物，包括大群的魚類、不同種類的底棲無脊椎生物及吸引了一些哺乳類動物在河邊濕地聚居。

香港河溪的生物很受季候雨所帶來深遠的影響。這些季候雨會增加水流的速度透過漂流和浸蝕從而減低了動物和海藻的的數量，除此之外、水底的動物和牠們所居住的沉積物有一個很強烈的關係；流水對溪澗河床的顆粒結構起著決定性的因素，從而亦影響溪澗裏生物的分佈和數量。

溪澗不同區域的河床，其顆粒大小視乎水流動力的大小。在淺流區因水流較急速，河床主要由大石和卵石組成；而水潭及水流緩慢之處則見大量幼小顆粒如粉砂、泥土和幼沙。

海岸線的主要物種

香港水域內共有二百多個島嶼，不同的海洋或海岸生物棲息於不同類型的海岸；本港水域的魚類、甲殼類、軟體動物及其他海洋生物。能夠令香港擁有這麼多樣性的海洋生物，最主要是香港位於亞熱帶地區，海洋環境適合熱帶和溫帶氣候的動物生長。香港位於珠江河口的東面。珠江流出的淡水主要流入香港西面水域，東面水域則以海洋特性為主，受珠江河水的影響甚微，因而擁有這種特別的水文狀況。

植被

我們可在香港曲折的海岸線找到許多以不同方法去適應海岸環境的植物：
一些固沙的植物，如海濱花后 (*Imopoea pes-caprae*)、海埔姜 (*Vitex rotundifolia*) 和蟛蜞菊 (*Wedelia prostrate*)，而在沙灘後方可找到苦郎樹 (*Clerodendrum inerme*)、露兜樹 (*Pandanus tectorius*) 和草海桐 (*Scaevola sericea*)。再往後方則可找到幾種陸地常見的灌木或樹種如黃槿 (*Hibiscus tiliaceus*)，血桐 (*Macaranga tanarius*) 和海芒果 (*Cerbera manghas*)。

匍匐植物如天門冬 (*A.cochinchinensis*)、雞眼藤 (*M.parvifolia*)、海刀豆 (*C.maritima*)、厚藤 (*I. pes-carprae*)、及灌木如草海桐 (*Scaevola sericea*)、濱排草 (*Lysimachia mauritiana*)、海埔姜 (*Vitex rotundifolia*) 和台灣海棗 (*Phoenix hanceana*) 常可在岩岸的後方找到。

濺水帶長滿地衣，潮澗帶高至中的界線長滿藻青菌 (*Kyrtuthrix maculans*)，中間則長滿覆蓋性藻類如褐殼藻 (*R. expansa*)；半葉馬尾藻 (*S. hemiphyllum*) 則常見於潮下帶。

在泥灘上，可以找到老鼠簕 (*Acanthus ilicifolius*)、桐花樹 (*Aegiceras corniculatum*)、海茄苳 (*Avicennia marina*) 和秋茄 (*Kandelia obovata*) 這些香港常見的紅樹品種。

動物

以國際標準而言，品種可算繁多。香港最豐富的珊瑚群落位於東面水域，擁有約80種石珊瑚及至少14種軟珊瑚，這些珊瑚直接附生在岩岸淺水區上的岩石和石床。我們可以找到看起來像枝節以實質是整個珊瑚蟲的群體，這些小動物屬刺胞動物門，牠們擁有一副堅硬的骨骼而柔軟的身體，牠們通常依附在岩石上或是已經死去珊瑚蟲的骨骼上。在珊瑚上看起來像花朵的組織，其實是珊瑚的觸鬚用作水中覓食。

身處濺水帶，你可以在隨處找到一些在潮汐間帶最強壯的小動物如草食性海螺和海蟑螂 (*Ligia exotica*)，草食性帽貝 (*Collisella dorsuosa*) 與濾食性藤壺 (*Chthamalus malayensis*) 就聚居在海浪沖刷程度較高的岩岸上。當潮水升高時，草食性帽貝會出來吃海藻；而潮水退下時，牠們便會回到棲息之處。

石鼈 (*Acanthopleura japonica*) 是高度耐旱因而極適合於岩岸生活的動物，石鼈是高效率的草食動物，有能力挖取覆蓋性藻類和微藻類生物膜作為食物，牠們與鱗笠藤壺 (*Tetraclita japonica*, *T. squamosa*) 和大藤壺 (*Megabalanus volcanoare*) 在不同區域的潮汐間帶隨處可以找到。

貽貝和海膽在半空曠的岸邊比比皆是。海膽與貽貝居住一起是可以從捕食者中自衛和免被大浪沖走。

此外，也有多種海魚在本港水域繁殖。早春時分，大鵬灣沿岸便見鯛魚其中一種——真鯛的大量魚苗出現，牠們是香港東面水域其中一種最典型的海洋生物。此外你亦可發現在香港最常見的魚類——賴氏犁齒魚 (*Entomacrodus stellifer lighti*) 岩岸中出現。

綠海龜 (*Chelonia mydas*)於南丫島深灣沙灘產卵。並全年都有兩種海洋哺乳類動物出沒，其中較廣為人知的是印度太平洋駝背豚 (*Sousa chinensis*)，俗稱中華白海豚，另一種是江豚 (*Neophocaena phocaenoides*)。

中華白海豚喜愛河口環境，在香港西面水域棲息，當地是珠江河將淡水流入海中的地方；江豚則在東面及南面水域棲息，這些水域以海洋特性為主。

河溪的主要物種

河溪是水流的生態系統，流水往往沿同一方向從源頭 (起點) 流入河口 (注入其他水體)。它們嵌於陸地之間，因此與沿岸的地點有不可分割的關係。由於流水所經的土地差異可以很大，河溪的源頭和河口經常展現很大的差別。本地河溪的生態價值很高，經常可找到各類水生植物、昆蟲、螺貝、甲類動物、淡水魚、兩棲類和爬蟲類。*Marcopodus hongkongensis* 與香港蠑螈 (*P. hongkongensis*) 都是香港的原生品種。

植被

河溪植物是指生長於河溪生境裏或旁邊的淡水植物，某些河溪植物主要生長在水體的邊緣，而其他則在淺水區長出及在經常浸在或浮於較深水處。

長於河溪邊緣的植物大多並不是水生植物，而是一些需要濕潤環境或長期水的供應才可生長的灌木和樹木，如榕屬(牛乳樹*F. fistulosa*, 對葉榕*F. hispida*和舶梨榕*F. pyriformis*)。

一些根部生在水裏、經常浸在水裏或漂浮帶葉植物各以不同的適應方法長於流水間。絲狀藻(*Cladophora* spp.)、蘚類(*Fissidens* 和 *Clapopodium* spp.)和苔類常見於山澗一簇簇地生長。

生物

香港河溪的主要生物是昆蟲，但亦可找到甲殼動物、軟體動物、兩棲類及少量的魚。所有河溪昆蟲都是水陸兩生型，幼蟲階段在水中，而成蟲則是陸棲的。由於水溫穩定，許多昆蟲一年四季都會進行繁殖，包括蜉蝣、石蠅 (績翅目) 如*Kamimuria.* Spp、石蛾(毛翅目)如*Ganoema extensum*及許多蜻蜓和豆娘。豆娘(*Euphaea decorate* and *Rhinocypha perforata*)和最常見的池耶(*P. tigrina*)都是香港山溪常見的昆蟲。

其他小生物如螺貝(*Brotia hainanensis*)、雙殼類 - 河蜆(*Corbicula fluminea*)和大顎蜘蛛 (*Tetragnathidae mandibulata*) 有修長的身體和長而幼細的腿以便在大石上移動。在香港河溪不但數量多而且分佈廣泛。淡水蟹 (*Cryptopotamon anacoluthon*) 和大閘蟹 (*Eriocheir japonicus*)常藏身於河岸的石塊下。若你仔細找，可能會在河床找到知名的兩棲類 - 香港蠑螈(*Paramesotriton hongkongensis*)和瀕危品種盧文氏小樹蛙。細小的魚類如*Parazacco spilurus, Pseudogastromyzon myers*和*Ctenogobius duospilus* 在香港的數量多，且分佈廣泛。香港淡水最常見和分佈最廣泛的蝦類 - 虎紋蝦(*Caridina cantonensis*) 和 *Macrobrachium hainanense* 在水潭的枯葉上常可找到。

海洋環境保育

香港經濟發展及人口不斷增加，透過人類使用如河口疏導及築堤限制了海洋生物的水域，香港的海岸及各種海洋生物正飽受威脅，很多海洋中的動植物因而被淘汰。海岸公園及海岸保護區規例是一個保育計劃落實保護現存的海洋環境。

於1995年頒布的《海岸公園條例》(第476章)是為了香港海洋存護，該條例賦予漁護署劃定、管理及管轄香港海岸公園及海岸保護區。於1996年7月頒布的《海岸公園及海岸保護區規例》訂定條文，禁止或管制在海岸公園及海岸保護區內進行某些活動。

海岸公園所佔的海域較大，可作為存護和康樂用途。相對之下，海岸保護區面積較小，但存護價值卻甚高，故主要用作存護、科學研究及教育用途，其管制也會較海岸公園嚴格。

在海岸公園及海岸保護區內，禁止進行以下活動 (根據漁護署〈海岸情報公園篇〉2003年)：

- 未經許可捕魚、捕獵、收集或擁有野生動植物或其部分
- 干擾受保護的海洋生物的卵、幼雛及哺育場
- 持有拖網、魚槍、炸藥或可用作捕魚的化學劑或電殛裝置
- 滑水、高速滑水或駕駛水上電單車
- 以超逾10海里的速度駕駛船艇
- 破壞沿岸任何海灘、泥灘、懸崖及海床
- 毀壞政府告示、標記等裝置
- 阻塞水道或污染海水或拋棄垃圾
- 未經許可進行商業活動、團體活動、繫泊及碇泊
- 海產養殖

此外，海岸保護區內亦禁止游泳、潛水及任何船艇活動。

香港的考察地點

直至現在，本港共設立四個海岸公園及一個海岸保護區：

- 海下灣海岸公園、印洲塘海岸公園及鶴咀海岸保護區於1996年7月5日成立
- 沙洲及龍鼓洲海岸公園亦於1996年11月22日成立
- 東平洲海岸公園則於2001年11月16日成立

我們建議同學可到下列地點進行考察：東平洲海岸公園、海下灣海岸公園及大蠔河。

1) 東平洲海岸公園

東平洲海岸公園是香港第四個海岸公園，於2001年11月16日指定。該海岸公園約佔270公頃海域面積，位於香港東北水域的大鵬灣，環繞平洲而設。珊瑚群落、海藻床和岩岸是東平洲重要的海洋生境。

東平洲海岸公園擁有多樣化的海洋生態。位於大塘灣及亞媽灣，共錄得超過四十五種石珊瑚及超過一百三十種珊瑚魚類例如蝶魚，雜衣，石班，火點，細鱗以及沙鯭。同時，這裏亦錄得超 過一百種海洋無脊椎動物。現時當局在大塘灣及亞媽灣劃定兩個核心區，以加強保護及維持多種重要的生物。

東平洲海岸公園找到四十多種海藻。在龍落水的沿岸地帶，長滿棕色(*Sargassum* sp.)、紅色(*Hypnea* sp.)及綠色(*Ulva* sp.)的海藻床，被認為是香港海藻床之冠。

這裡正好提供學習岩石形成和研究潮汐間帶生態的機會。在岸上游，你會發現地衣，覆蓋性藻類與海螺，與之同時在潮間帶可以找到鱗笠藤壺 (*Tetraclita japonica*), 帽貝，鐘螺，石鱉和胎貝 (*Perna viridis*)。

交通：渡輪於星期六日及公眾假期由馬料水碼頭(鄰近大學火車站)開往東平洲碼頭。

建議時間：六小時

2) 海下灣海岸公園

海下灣海岸公園是一個受遮蔽的海灣，水質潔淨，因此為各種海洋生物提供了良好的海洋棲居環境。海岸公園位於西貢西郊野公園北端，是一個受遮蔽的海灣。其海上界線以連接嚮螺角與棺材角尖端的直線為限，並穿越銀洲和磨洲的北端。

此海岸公園海域面積約佔260公頃眾所周知，海下灣海岸公園有繁茂的珊瑚群落。本港錄得的八十八種石珊瑚中，有五十九種可在該處找到。多個石珊瑚品種，包括刺星珊瑚、刺葉珊瑚、蜂巢珊瑚、角蜂巢珊瑚、角孔珊瑚、小星珊瑚、石葉珊瑚、牡丹珊瑚及扁腦珊瑚，是海岸公園內的一個或多個珊瑚區內的優勢品種。

該海岸公園已錄得超過一百二十種珊瑚礁魚類例如石班、火點和雜衣。此外，還找到多種海洋生物如海星及水母，亦進一步證明該處的生態價值極高。

海下灣海岸公園的紅樹林可找到幾種真紅樹品種，包括秋茄、桐花樹、海漆、木欖、海欖和欖李。紅樹林地區常用作幼魚、潮間及潮下帶無脊椎動物的哺育場。

交通：可乘搭7號專線小巴從西貢市到海下村或乘搭九巴94號從西貢碼頭往黃石碼頭，於高塘下車，沿海下路步行約一小時便可到達。

建議時間：六小時

3) 大蠔河

大蠔位於大嶼山東北部，擁有多樣化的生態環境，包括風水林、灌木林、草地、溪流、季候性沼澤蝙蝠的棲息地在地勢較高的山坡；而大蠔河低處可找到紅樹林、海草床和泥灘。

大蠔河擁有全港最豐富的淡水魚品種，共錄得五十三種，具重要的存護價值，因而大蠔河已在1999年列為「具特殊科學價值地點」。此外，這裡亦可找到十種兩棲動物，其中香港特有的盧氏小樹蛙 (*P. romeri*) 亦於這裡棲息。大蠔西岸仍可找到稀有的貝克鹽藻 (*H. beccarii*)，扶養本地稀有的馬蹄蟹。

在河流上空同樣生機處處。大蠔河口便可以發現小白鷺，八哥，白胸翡翠與白鶺鴒。河流兩旁孕育無數的有花植物，在花叢中飛躍的蝴蝶與飛蛾隨處可見。

交通：可乘搭新大嶼山巴士36號 (循環線) 由東涌市中心前往小蠔灣，車程約五分鍾，從翔東路下車。班次為：7:45am, 10:00am, 3:00pm, 5:00pm 及 7:30pm

建議時間：三小時

河溪/海岸線考察守則

- 切勿翻起海岸石頭
- 切勿捕獵或騷擾海洋生物
- 切勿亂拋垃圾
- 切勿污染海水
- 愛護海洋及水環境

河溪／海岸線考察工作紙

1) 考察地點：________________ 公頃

2) 列出三種於該處找到的動物：

______________________ 辨別特徵：____________________________________

______________________ 辨別特徵：____________________________________

______________________ 辨別特徵：____________________________________

3) 列出三種於該處找到的植物：

______________________ 辨別特徵：____________________________________

______________________ 辨別特徵：____________________________________

______________________ 辨別特徵：____________________________________

4) 寫出河溪及海岸線遇到的困難：

- ____
- ____
- ____

5) 動物如適應潮澗帶或河溪惡劣的環境？

- ____
- ____

6) 請建議一些保護河溪及海岸線的一些方法：

- ____

4. Field Trips Applying Environmental Principles

This section deals with **field visits applying environmental principles**, such as EPD Resources Centres, China Light and Power Company, Fung Yuen Butterfly Reserve, etc. These sites are categorized by the environmental principles they apply to. By visiting these atypical outdoor classrooms, students can practice the environmental principles taught by the textbook and CD-Rom in the education packages.

Visiting Lions Nature Education Centre

The field visits applying environmental principles include:

Environmental Principles Applied	Recommended Field Sites
1. Environmental Awareness	• EPD Environmental Resource Centres • Lions Nature Education Centre
2. Corporate Environmental Management	• Cathy Pacific City • China Light and Power Company
3. Sustainable Development	• HK Planning and Infrastructure Exhibition Gallery • Hong Kong Airport Core Programme Exhibition Centre
4. Resources and Waste Management	• Landfill Sites (SENT) • Water Treatment Works in Shatin
5. Food Safety	• Various supermarkets
6. Organic Farming	• Kadoorie Farm and Botanical Garden (KFBG) • Produce Green Foundation Organic Farm
7. Ecotourism	• Fung Yuen Butterfly Reserve • Ma Shi Chau Special Area

4. 應用環境法則的野外研習

有關機場基建項目的展覽

本部分嘗試介紹一些應用環境法則的野外研習地點，例如：環保署環境資源中心、中華電力公司、鳳園蝴蝶保育區等。

以下研習地點是以不同環境法則而分類。希望透過親身參觀或探訪這些户外教室，學生可以將教科書或光碟中提及的環境法則，好好學以致用。

有關應用環境法則的參觀地點包括：

所應用的環境法則	建議的參觀地點
1. 環保意識	• 環境保護署環境資源中心 • 獅子會自然教育中心
2. 企業環境管理	• 國泰城 • 中華電力公司
3. 可持續發展	• 香港規劃及基建展覽館 • 機場核心計劃展覽中心
4. 資源與廢物管理	• 新界東南堆填區 • 沙田濾水廠
5. 食物安全	• 各超級市場
6. 有機耕作	• 嘉道理農場暨植物園 • 綠田園
7. 生態旅遊	• 鳳園蝴蝶保育區 • 馬屎洲特別地區

4.1 Applying Environmental Awareness

People of Hong Kong

Country parks

The problems of pollution in Hong Kong are similar to those encountered in most comparable urban areas elsewhere in the world. The government has put tremendous effort in tackling the problems and in promoting the issues to the public through many campaign activities and a series of publications.

The Environmental Campaign Committee (ECC), along with Environmental Protection Department (EPD), organizes events to promote environmental awareness and administers part of the Environment and Conservation Fund.

The Agriculture, Fisheries and Conservation Department (AFCD) also contributes in environmental awareness by establishing country parks, marine parks and nature reserves.

In order to raise environmental awareness, students should visit **EPD's Environmental Resource Centres** and **AFCD's Lions Nature Education Centre**.

1) EPD's Environmental Resource Centres

The Environmental Resource Centres aim to provide information about the environment to the community. The first centre opened in **Wan Chai** in 1993 and was followed by the establishment of centre in **Tsuen Wan** in 1997. They feature environmental activities, games and general information for the public, and guided tours to be booked in advance.

Display inside Wanchai centre

A **Mobile Environmental Resource Centre** started operating in 2001, featuring all of the activities and resources available in the fixed centres. It is the first liquefied petroleum gas van and also has room for a "green desk" where visitors can obtain specific information for school projects, home management and other uses, and have their questions answered.

Entrance of Tsuen Wan centre

Located in Luen Wo Hui, the latest **Fanling Environmental Resource Centre** opened in 2003 and is the largest centre yet. About 400 square metres in size, the centre is fitted out with furniture mostly made from recyclable materials and has a mural at its entrance that was created from waste construction materials. Among its many features are interactive games, models on how to reduce pollution, demonstrations of noise pollution, a fuel cell battery, and a solar energy display.

Fanling centre

Using the Fanling Environmental Resource Centre as an example, the typical centre features the following:
Exhibition Halls -

- Clean Air
- World of Silence (learn noise pollution)
- Reduce Waste, Start from Me
- Protect Our Water Resources
- Environmental Planning and Assessment
- Community Education (Computer Gallery)
- Global Environmental Issues

Other Services and Facilities -

- Library (collection of green publications)
- 50-people Lecture Room (for workshops)
- Guided Visits for schools and community groups

One showroom display

Interactive feature for visitors

Enquiries and booking:
Wanchai: 2893 2856
Tsuen Wan: 2944 8204
Fanling: 2600 4016
Website: http://www.epd.gov.hk/

Addresses and Transportation:
Wanchai Centre -
221 Queen's Road East, Wanchai, HK (Walk for 10 minutes from Wanchai MTR station towards Wu Chung House)
Tsuen Wan Centre -
Tak Wah Park, Tak Wah Street, Tsuen Wan, NT. (Just a 5-min walk from Tsuen Wan MTR station through Tsuen Wan Town Plaza)
Fanling Centre -
2/F., Government Accommodations, Grand Regentville, 9 Wo Mun Street, Luen Wo Hui, Fanling, NT. (Take 278K Bus to Luen Wo Hui from Fanling KCR station)

Notes on field visits:

- The Resource Centres open from 10:00am to 5:00pm on weekdays and Saturdays. No admission fee is required.
- Guided visits and environmental education programs for schools and community groups are offered through booking in advance.
- Do not damage or take away any objects.
- Listen and follow the advice of local management authority.
- **Suggested time**: 2-3 hours at the centre

2) AFCD's Lions Nature Education Centre

Outlook of the Centre

Shell house

Nature trail

Dragonfly pond

Lions Nature Education Centre, of 16.5 hectares in size, is located in the Tsiu Hang Special Area in Sai Kung. It was formerly a government experimental farm established in 1950s. Because of its conservation value and potential for nature education, the site was designated under the Country Parks Ordinance as a special area in 1987. Sponsored by the Lions Clubs, it was further developed into an education center in 1991.

The Centre has five exhibition halls, namely Countryside Hall, Agriculture Hall, Fisheries Hall, Insectarium and Shell House, and over 17 outdoor exhibition corners.

Just a few highlights on the exhibition spots:

- **Countryside Hall** introduces the biodiversity in mangrove habitats and country parks as well as marine conservation.
- **Agriculture Hall** illustrates the agricultural development and use of biotechnology in the industry.
- **Insectarium** teaches classification, metamorphosis, zoogeographical distribution of local insect species.
- **Dragonfly Pond** hosts about 52 dragonfly species, about one quarter of total HK species.
- **Shell House** displays more than 5000 pieces of beautiful shells, promoting marine conservation.
- **Arboretum**, of 0.5 hectare in area, has collections of local Camellias and Azalea, typical plants of conifers and "Fung Shui" woodland.
- **Tsiu Hang Nature Trail** is a 650-metre circular path showing many interesting traits of nature.

Fisheries gallery display (12)

Notes on field visits:

- Address: Hiram's Highway, Tsiu Hang, Sai Kung.
- You may arrange a group of 30 persons or above for group visits. Make telephone reservation first (Tel: 2792 2234) and then provide written confirmation.
- Opening hours: 10:00 – 16:00 (close on Tuesdays)
- Bring more water in summer, and also meals, hats and towels. Long sleeved clothing and trousers are recommended to avoid insect and mosquito bites.
- Listen and follow the advice of local management authority. You may also follow the practical worksheet on the next page.

Transportation: Take KMB Bus No. 92 from Diamond Hill MTR station to Sai Kung, and get off at Pak Kong stop and walk to the Education Centre.
Suggested time: 3–5 hours

Practical Worksheet:
- for visiting Lion's Nature Education Centre

1) Name 3 herb species of Chinese medicine and their values:

   ____________________ values: ________________________________

   ____________________ values: ________________________________

   ____________________ values: ________________________________

2) Apart from its beauty, what are other functions of shells?

   ____

3) There are about _______ dragonfly species in the "Dragonfly Pond",

   representing ____________% of total species in Hong Kong.

4) "Fisheries Gallery" introduces fisheries development in Hong Kong. Which is the most effective fishing method?

   ____

   Artificial reef functions as ____________________________________

5) "Insect Gallery" demonstrates the life cycle of insects. How do butterflies adapt to the living environment?

   - __________________________ • __________________________
   - __________________________ • __________________________

6) Other questions/comments:

   ____

   ____

4.1 應用環保意識

正如世界各地許多高度發展的城市，香港亦面對各種污染問題，政府已就此推行周詳的解決方案，而這些正在實施的計劃已在多方面取得成效，如污水處理及廢物棄置。當局每年都會出版刊物「**香港環境保護**」，講述香港環境的最新狀況。

環境保護運動委員會(ECC)與環境保護署(EPD)攜手舉辦多項活動以提高公眾的環保意識，並負責管理部分環境及自然保育基金。

漁農自然護理署(AFCD)亦透過成立郊野公園、海岸公園及自然保育區，推廣環保意識。

同學可參觀環境保護署轄下的**環境資源中心**及漁農自然護理署的**獅子會自然教育中心**，以取得更多有關環境保護的資料。

1) 環境保護署環境資源中心

環境資源中心旨在為公眾提供有關環境的資訊。首間環境資源中心於一九九三年十二月在**灣仔**開幕，隨後**荃灣**環境資源中心亦於一九九七年成立。中心舉辦有關環境保護的活動和遊戲，並提供相關資訊，現已深受市民歡迎，團體預約參觀經常在個多月前已滿。

流動環境資源中心於2001年開始運作，與其他資源中心一樣，會舉辦有關環境教育的活動和提供相關資訊。這亦是首輛石油氣貨車，車上設有環保詢問處讓學生、主婦及居民獲取有關專題研習、家居管理及其他資訊，並回答他們的查詢。

注意事項：
- 資源中心平日及星期六的開放時間為早上十時至下午五時，費用全免。
- 中心可提供免費導賞服務及環境教育計劃予已經預約的學校及社區團體。
- 不可損壞或帶走任何物品。
- 聽從及遵守當值人員的指示。
- **建議時間：**每間中心約2–3小時
- 查詢及預約：
  灣仔：2893 2856
  荃灣：2944 8204
  粉嶺：2600 4016

地址及交通：

灣仔環境資源中心–
香港灣仔皇后大道東221號
(由灣仔地鐵站向胡忠大廈步行10分鍾便可到達)

荃灣環境資源中心-
新界荃灣德華街德華公園
(由荃灣地鐵站經荃灣城市廣場步行5分鍾便可到達)

粉嶺環境資源中心-
新界粉嶺聯和墟和滿街9號御庭軒政府綜合大樓2樓
(到粉嶺火車站乘搭278K巴士到聯和墟)

2) 獅子會自然教育中心

獅子會自然教育中心位於西貢蕉坑特別地區，佔地16.5公頃，前身為政府於1950年代成立的實驗農場。鑑於該區具有保育價值及可用作自然教育中心，政府於1987年根據郊野公園條例將這裡劃為特別地區，其後獲得獅子會的贊助，將這裡發展為戶外教育中心，並於1991年開幕。

中心設有五個展覽館，包括郊野館、農館、漁館、昆蟲館及貝類館，戶外另設有超過17個展覽區。參觀人數如有30人或以上，可致電24229431向中心預約導賞服務。

展覽館以不同主題劃分，內容豐富：
- **郊野館** 介紹紅樹林生境和郊野公園的生物多樣性以及海岸保育。
- **農館** 解釋農業發展及該行業應用的生物科技。
- **昆蟲館** 介紹昆蟲的分類、變態現象及本地品種的分佈。
- **蜻蜓池** 共可找到52種蜻蜓，相等於本港發現的蜻蜓品種數目四分之一。
- **貝殼館** 展出約5000枚美麗的貝殼，同時藉此宣傳海岸保育。
- **標本林** 面積約半公頃，收集了本地茶花和杜鵑及風水林和針葉樹的代表品種。
- **蕉坑自然教育徑** 是一條循環路徑，長650米，沿途展示中心所介紹的有趣事物。

注意事項：
- 地址：西貢蕉坑西貢公路
- 參觀人數如有30人或以上，可致電2792 2234預約團體參觀及遞交書面申請。
- 開放時間：10:00–16:00(星期二休館)
- 請穿著輕便及合適衣服。
- 夏天應帶備充足的水，亦要帶食物、手巾和帽。穿著長袖衣服和長褲可避免蚊叮蟲咬。
- 聽從在場職員的指示，你亦可參考本章的工作紙。

交通：於鑽石山地鐵站乘搭92號巴士往西貢，於北港站下車，步行往教育中心。
建議時間：3–5小時

工作紙：

－ 參觀獅子會自然教育中心

1) 試舉3種「中草藥園」裡的草藥及其藥用價值：

_______________ 藥用價值：_______________________________

_______________ 藥用價值：_______________________________

_______________ 藥用價值：_______________________________

2) 除美觀價值外，貝殼還有什麼重要性？

____

3) 「蜻蜓池」中約有 __________ 種蜻蜓，相當於全港可發現蜻蜓品種的

__________ %

4) 「漁館」介紹本港漁業發展，當中最有效的捕魚方法是什麼？

____

人工漁礁的作用是 _______________________________

5) 「昆蟲館」內有各樣昆蟲的介紹，當中蝴蝶有什麼適應環境的方法？

- ____________________ • ____________________
- ____________________ • ____________________

6) 其他問題／意見：

____

4.2 Applying Corporate Environmental Management

Hong Kong

Hong Kong is well known for its free trade and economic development. Local and foreign investors have established business in different sectors employing thousands of workers. While making profits is still the main objective, many corporations recognize that their operations have environmental impacts on the society along with the growth of business.

Nowadays, corporations pay more attention on social responsibility such as energy conservation and environmental protection. Not only they practice the policies of environmental management within the companies, some of them are willing to demonstrate the practices to the public using showrooms and exhibitions.

← Companies should practice social responsibility

To explore the concept of corporate environmental management, it is suggested to visit **Cathy Pacific City** at the Airport, and **China Light and Power Company** in Tuen Mun.

1) Cathy Pacific City

Cathay Pacific is an international airline in Hong Kong, offering scheduled passenger and cargo services to all over the world. Cathay Pacific City, at the southeast corner of Hong Kong International Airport in Lantau Island, hosts over 3,000 staff with the necessary facilities. Opened in 1998, the premises spans a total of 1.3 million square feet, making it one of the largest corporate headquarters in Asia.

Inside the premises

The complex's three office towers illustrates a new standard in ergonomic, open-plan office design, offering natural lighting and panoramic views. Other basic facilities for the staff include a self-learning center, a 23-floor hotel, an auditorium, a weight training room, etc.

Natural interior lighting

With a nature-oriented design and using environmentally friendly materials throughout the construction, the complex uses an intelligent *Building Management System* with controls interior lighting and temperature to ensure optimum use of energy. The main office facilities are equipped with a high central atrium, with large glazed windows designed to absorb outside natural lighting.

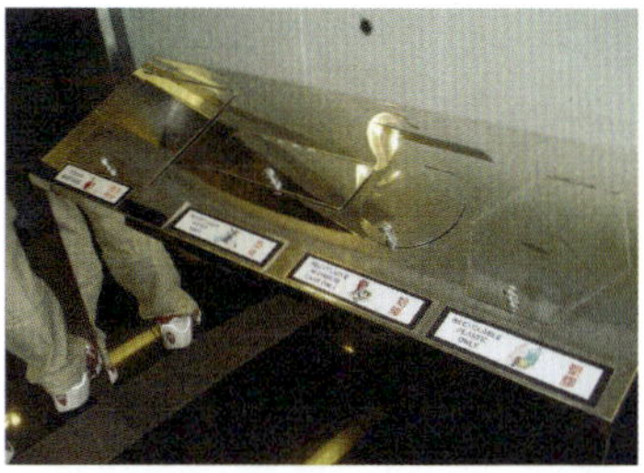
Recycling bins everywhere

The City's other green practices include:

- A comprehensive recycling system (at the lift corridor of each floor) for collecting waste paper, aluminum cans and plastic bottles.
- Lots of signs and notices to remind staff and visitors to practice energy conservation.
- Lots of trees have been planted to keep a comfortable, green environment, providing natural fresh air inside the building.

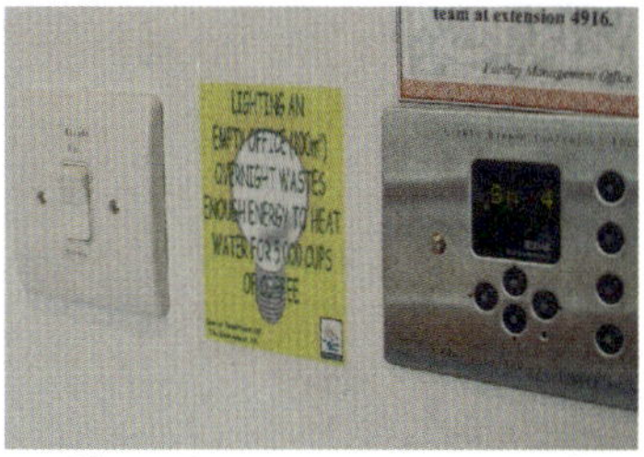
Signs for energy conservation

Notes on field visits:

- Make prior reservation with Cathay Pacific. The guided tour is about 2 hours.
- Put on suitable clothes and footwear.
- Do not damage or take away any objects from the premises.
- Listen and follow the advice of local management authority. You may also follow the **practical worksheet** at the end of chapter.
- Pay attention to personal safety during the visit.

Staff explains environmental management to students

Transportation: Take the S64 Bus from Tung Chung MTR station towards the Chek Lap Kok Pier. Cathay Pacific City is located opposite to Super Terminal I.

Suggested time: 2 hours at the site.

2) China Light and Power Company (ElectriCity)

Entrance of exhibition and display centre

The China Light and Power (CLP) Group generates and supplies electricity to over 5 million people living in Kowloon, New Territories and over 100 outlying islands in Hong Kong. The Group operates power stations at Castle Peak, Black Point and Penny's Bay. It has developed an Environmental Management System (EMS) that complies with international **ISO 14001** certification, providing a robust structure for addressing environmental issues in all aspects.

Electricity generator

Interactive displays

Control Panel

Some of the company's green measures include:

- Reducing energy consumption by improving efficiency.
- Promoting the use of renewable energy such as solar, wind and wave power.
- Reducing atmospheric emission of harmful pollutants.

CLP has set up an exhibition and display center called **ElectriCity** located at Castle Peak in Tuen Mun, allowing visitors to better appreciate the basic knowledge of electricity and to experience the latest technologies of the power industry. It aims to demonstrate how electricity is generated and supplied to the homes of millions through this educational facility.

The entire exhibition centre has four major galleries:

1. CLP History and Overview – shows the development of Castle Peak Power Company.
2. Power Generation Area – teaches how to generate electricity and its basic principles.
3. Transmission and Distribution Area – demonstrates how electricity is transmitted and distributed to the homes of the public.
4. Safety, Health & Environment Culture Area – let people understand how the measures of these aspects are incorporated in the business.

Notes on field visits:

- The exhibition gallery opens from Monday to Friday (9am to 5pm) and only accepts group visitors. Make prior reservation with CLP at 2678 4001 or email to gbgpa@clp.com.hk.
- Do not damage or take away any objects from the premises.
- Listen and follow the advice of local management authority.
- Pay attention to personal safety during the visit.

Transportation: There is no direct public transportation to Castle Peak Power Station. Visitors must arrange for their own transportation to Lung Fu Road of Lung Kwu Tan or travel by taxi from Tuen Mun.
Suggested time: 3 hours at the exhibition centre.

Staff explaining environmental management to visitors

Practical Worksheet
- for visiting Cathay Pacific City

1) Size of Cathay Pacific City: ____________ hectares

2) Environmental problems encountered by the Company:

   Problem 1: ____

   Problem 2: ____

   Problem 3: ____

   Problem 4: ____

3) What are the solutions to these problems?

   Solution to Problem 1: ____________________________________

   Solution to Problem 2: ____________________________________

   Solution to Problem 3: ____________________________________

   Solution to Problem 4: ____________________________________

4) As an airline company, Cathay Pacific generates huge amount of waste from the airplanes everyday. How does the company tackle this problem?

   ____

   ____

   ____

5) Other comments/suggestions: ______________________________

   ____

4.2 應用企業環境管理

香港以其自由貿易和經濟發展聞名，本地和外國投資者於各行各營商，僱用了數以千計的僱員。雖然賺取利潤仍是商家的主要目標，但許多商業機構現已明白到其商業活動不單對社會的環境構成壓力，亦對其業務增長有所影響。

企業現時更關注節約能源和環境保護等社會責任，有些公司不但推行環境管理政策，更樂於將有關成果以陳列室或展覽形式向公眾展示。

為研習企業環境管理的概念，我們建議同學參觀分別位於大嶼山赤鱲角機場的**國泰城**及屯門龍鼓灘的**中華電力公司**。

1) 國泰城

國泰航空有限公司是香港一間國際航空公司，為全球各國提供定期的客運和貨運服務。國泰城位於大嶼山香港國際機場的東南方，於1998年啟用，為超過3,000名職員提供所需設施，佔地一百三十萬平方呎，為亞洲最大的企業總部之一。

城內三座辦公大樓採用了最新的人體工學標準和開放式辦公室設計，為室內提供自然光和全景景觀。其他為員工而設的基本設施包括自學中心、二十三層的酒店、演講廳及健身室等。

國泰城在設計時已考慮到環保問題，因此整項建築工程均採用環保物料，並以電腦化樓宇管理系統控制室內照明和溫度，以確保適當地使用能源。辦公室的主要設施均圍繞中央的中庭而建，並透過安裝玻璃來吸收自然光。

其他環保措施包括：
- 於每層升降機的走廊放置回收箱以收集廢紙、鋁和膠樽。
- 張貼提示和告示提醒員工和訪客節約能源。
- 種植樹木美化環境，並為室內提供新鮮空氣。

注意事項：
- 參觀國泰城需要預先預約，導賞團需時太約兩小時。
- 穿著合適衣服及鞋。
- 不可損壞或帶走國泰城內的任何物件。
- 聽從及遵守在場職員的指示，你亦可參考本章的工作紙。
- 注意個人安全。

交通： 由東涌地鐵站乘搭S64路線往赤鱲角碼頭，國泰城位於超級1號貨站對面。

建議時間： 約2小時

2) 中華電力公司（「電力世界」）

中華電力公司(CLP)主要為九龍、新界及超過100個離島的居民發電及供電，總人口超過五百萬，其三間發電廠分別位於青山、龍鼓灘和竹篙灣。中電建立了一套環境管理系統(EMS)，並獲頒國際認可證書**ISO 14001**，以提供一套完善的方案處理各種環境問題。

中電採用的其他措施包括：
- 提高效率以減少能源消耗
- 提倡應用可再生能源，如太陽能、風能和水波動能
- 減少廢氣排放量

中電於屯門青山發電廠內建立了一個展覽及陳列中心，名為「電力世界」，讓訪客認識電力的基本原理、親身體驗最新的電力科技及了解電力的生產及輸送過程。

整個展覽中心共分為四個主要部分：
1. 中電歷史及發展 - 展示青山發電廠的發展。
2. 發電 - 展示如何發電及其基本原理。
3. 輸電及配電 - 展示電力如何輸送及分配給用戶。
4. 安全、健康及環保 - 讓你認識中電如何在發電及輸配電的過程中推行有關的措施。

注意事項：
- 展覽館於星期一至五上午九時至下午五時開放，但只招待團體參觀。請先致電 2678 4001 或電郵至 gbgpa@clp.com.hk 預約參觀。
- 穿著合適衣服及鞋。
- 不可損壞或帶走電力世界內的任何物件。
- 聽從及遵守在場職員的指示，你亦可參考本章的實用工作紙。
- 注意個人安全。

交通：沒有公共交通直達青山發電廠，訪客需自行安排交通到龍鼓灘龍富路，或由屯門乘搭的士前往。

建議時間：約3小時

工作紙：
－ 參觀國泰城

1) 國泰城的面積：＿＿＿＿＿＿＿＿ 公頃

2) 國泰航空所面對的環境問題：

- ＿＿＿＿＿＿＿＿＿＿＿＿＿＿＿＿＿＿＿＿
- ＿＿＿＿＿＿＿＿＿＿＿＿＿＿＿＿＿＿＿＿
- ＿＿＿＿＿＿＿＿＿＿＿＿＿＿＿＿＿＿＿＿
- ＿＿＿＿＿＿＿＿＿＿＿＿＿＿＿＿＿＿＿＿

3) 國泰航空怎樣解決上述問題?

問題一：＿＿＿＿＿＿＿＿ 解決方法：＿＿＿＿＿＿＿＿＿＿＿＿

問題二：＿＿＿＿＿＿＿＿ 解決方法：＿＿＿＿＿＿＿＿＿＿＿＿

問題三：＿＿＿＿＿＿＿＿ 解決方法：＿＿＿＿＿＿＿＿＿＿＿＿

問題四：＿＿＿＿＿＿＿＿ 解決方法：＿＿＿＿＿＿＿＿＿＿＿＿

4) 國泰航空是一間航空公司，每日都會產生大量廢物。國泰如何解決這個問題？

＿＿＿＿＿＿＿＿＿＿＿＿＿＿＿＿＿＿＿＿＿＿＿＿

＿＿＿＿＿＿＿＿＿＿＿＿＿＿＿＿＿＿＿＿＿＿＿＿

＿＿＿＿＿＿＿＿＿＿＿＿＿＿＿＿＿＿＿＿＿＿＿＿

5) 其他意見/建議：＿＿＿＿＿＿＿＿＿＿＿＿＿＿＿＿＿＿

＿＿＿＿＿＿＿＿＿＿＿＿＿＿＿＿＿＿＿＿＿＿＿＿

4.3 Applying Sustainable Development

The concept of sustainability has become a popular topic nowadays. In order to move towards sustainable development, it is necessary to make choices about how best to advance our economic and social development while protecting the natural environment. In simple terms, sustainable development for Hong Kong means:

- finding ways to increase prosperity and improve the quality of life while reducing overall pollution and waste;
- meeting our own needs and aspirations without making damage to the prospects of future generations; and
- reducing the environmental burden and helping to preserve common resources.

Urban development

For studying sustainable development, it is suggested to visit **Hong Kong Planning and Infrastructure Exhibition Gallery** in Central and **Hong Kong Airport Core Programme Exhibition Centre** in Ting Kau.

1) Hong Kong Planning and Infrastructure Exhibition Gallery

Entrance of the Gallery

Website:
www.info.gov.hk/infrastructuregallery

Located inside Edinburgh Place in Central, the **Hong Kong Planning and Infrastructure Exhibition Gallery** showcases the planning and infrastructure projects by the Hong Kong Government to enhance the quality of life, considering the three elements of sustainability (i.e. economics, environment and society).

The admission-free gallery comprises four main sections: Town Planning, Tourism, Transport and Logistics, and Environmental Protection.

Town Planning deals with the latest strategic planning and infrastructure development. The Gallery has details on shoreline enhancement of Victoria Harbor, Southeastern Kowloon development, renewal of older urban areas and new town development.

As **tourism** continues to feature prominently on the Government's agenda, the Gallery presents models and displays that illustrate the tourist kaleidoscope, West Kowloon cultural district, historical buildings as well as new tourist attractions.

Illustration of West Kowloon

As our partnership with the Pearl River Delta strengthens, so is the importance of **transport and logistics**. The Gallery demonstrates the future road and railway systems, and introduces some planned logistics facilities such as the 68-hectare Container Terminal 9 and a new logistics center at Chep Lap Kok for handling large sum of air cargo.

Harbourfront simulation

For **environmental protection**, there are many measures to bring about a higher quality and healthier living environment. One of the initiatives is the Harbour Area Treatment Scheme to improve water quality in Victoria Harbour. Another focus is "Cleaner Air" by reducing the concentrations of air particulates and nitrogen oxides at urban roadside.

Displays inside the Gallery

Last but not least, the Gallery has built an **Infrastructure Walk** capturing the major planning and infrastructure projects in unique linear presentation. This 18.5-metre walk is presented with special dramatic audio-visual effects using interactive features and hi-tech devices.

Transportation: Walk towards City Hall for 5 min from Central. The Gallery is located at the corner of ground level of City Hall. (Tel: 3101 6516)
Suggested time: 2 hours

Notes on field visits:

- Address: G/F, 3 Edinburgh Place, Central (next to City Hall).
- Opening hours are 10:00 to 18:00 on weekdays and public holidays but closed on Tuesdays.
- Use the **practical worksheet** at the end of this chapter.
- Guided tours provided by gallery staff are scheduled on every Saturday, Sunday and public holiday on 12:30 and 15:30 (45 minutes each tour).

Infrastructure walk

2) Hong Kong Airport Core Programme Exhibition Centre

Inside the Centre

The opening of the new Hong Kong International Airport at Chek Lap Kok in July 1998 marked the completion of the multi-project Airport Core Programme (ACP). The entire 10-project ACP links up the central business district on Hong Kong island to the new airport and Tung Chung New Town, bringing new transport infrastructure and providing new land for housing and commercial purposes.

Located in Ting Kau near Castle Peak Road, the **ACP Exhibition Centre** provides general information on the 10 core projects through the use of models, photographs, panel texts and videos.

Airport models

Exhibits of the Airport Core Programme

The core projects include:

- New Airport
- Airport Railway
- Lantau Link
- Western Harbour Crossing
- North Lantau Expressway
- West Kowloon Expressway
- West Kowloon Reclamation
- Route 3 Expressway (Tsing Yi to Kwai Chung)
- Tung Chung New Town development

The Centre is also an excellent spot to watch the Tsing Ma Bridge

Notes on field visits:
- Address: 401 Castle Peak Road, Ting Kau, NT. (Tel. 2491 9202)
- Opening hours are 10:00 to 17:00 everyday but closed on Mondays (except public holidays). Admission is free.
- Website: www.info.gov.hk/napco/

Transportation: From Tsuen Wan Ferry Pier, take KMB Bus either No. 234B towards Seacrest Villa or No. 53 towards Yuen Long (East). Get off after the Lido beach station.
Suggested time: 2 hours

Entrance of the Centre

Practical Worksheet
- for visiting Hong Kong Planning and Infrastructure Exhibition Gallery

1) List the displays inside the Gallery that fall into the following:

   Society: ______________________________

   Economics: ______________________________

   Environment: ______________________________

2) Town Planning – name some of the proposed developments in Victoria Harbour:

   - ______________________________
   - ______________________________
   - ______________________________

3) Tourism – would you think that historical, cultural buildings or the new attractions (such as Disneyland) attract more visitors to come to Hong Kong? Why?

   ____

   ____

4) Environmental Protection – What causes air pollution and the solutions?

   - ______________ solution: ______________
   - ______________ solution: ______________
   - ______________ solution: ______________

5) Which is the most attractive feature throughout the **Infrastructure Walk**?

   ____

   Reasons: ______________________________

4.3 應用可持續發展

社會現時越來越強調可持續的概念。在香港，**可持續發展委員會**現正就本港的可持續性鼓勵市民發表意見。為達致**可持續發展**，我們有需要就如何在滿足經濟和社會發展的需要和保育自然環境之間取得最佳平衡，作出知情的抉擇。簡言之，可持續發展對香港來說的涵義就是(資料來源：www.susdev.gov.hk)：

- 在追求經濟富裕、生活改善的同時，減少污染和浪費；
- 在滿足我們自己各種需要與期望的同時，不損害子孫後代的福祉；以及
- 減少對鄰近區域造成環保負擔，協力保護共同擁有的資源。

本部分主要探討可持續發展的原則，並以現代發展及建築如何平衡經濟、環境和社會這三方面為重點。在眾多考察地點中，我們建議參觀**香港規劃及基建展覽館**及**機場核心計劃展覽中心**。

1) 香港規劃及基建展覽館

香港規劃及基建展覽館位於中環愛丁堡廣場，陳列香港政府為改善市民生活，以及鞏固香港作為亞洲國際都會的地位而推行的連串規劃及基建計劃。

展館主要分四個部分：城市規劃、旅遊、交通物流和環境保護。

城市規劃是有關為改善香港生活環境而制定的策略性計劃和基建發展。這部分展示美化維港兩岸計劃、東南九龍發展、舊區重建及新市鎮發展。

政府致力推動本港的**旅遊業**發展，因此以模型和陳列品展示香港的旅遊景點、西九龍文娛藝術區、名勝古蹟及最新旅遊景點。

隨著香港與珠江三角洲加強合作，**運輸系統和物流**設施越趨重要。展覽館展示規劃中的道路和鐵路系統，介紹新的物流設施，如佔地68公頃的九號貨櫃碼頭及位於赤鱲角的新物流中心，以處理大量的航空貨運。

環境保護方面，政府為提高市民的生活質素和健康水平而推行了多項措施，其中包括為改善維多利亞港水質而推行的淨化海港計劃及減少市區路邊懸浮粒子和氮氧化物濃度的空氣清新計劃。

基建走廊全長18.5米，匯集多項主要規劃及基建項目。整個展覽配備特別的影視效果，以高科技和互動形式表達。影片長12分鐘，以粵語廣播，星期六及公眾假期則輔以普通話和英語廣播。

網址：www.info.gov.hk/infrastructuregallery

注意事項：

- 地址：香港中環愛丁堡廣場3號地下(大會堂旁邊)。
- 查詢電話：3101 6516
- **開放時間：** 每日上午十時至下午六時(公眾假期除外，逢星期二休館)免費入場
- 逢星期六、日及公眾假期下午12:30及15:30，館內職員會提供導賞服務。(每節為時45分鐘)。
- 聽從及遵守在場職員的指示。

交通： 由中環步行5分鐘到大會堂，展覽館設於大會堂側。
建議時間： 約2小時

2) 機場核心計劃展覽中心

新興建的香港國際機場位於赤鱲角，於一九九八年七月開幕，標誌著包括有多項工程的機場核心工程計劃完成。以新機場為主要工程的十項機場核心工程計劃，將港島的中環商業區和新機場以及東涌新市鎮連接起來，帶來新的交通基礎建設，並提供了大量的填海土地作房屋和商業用途。

機場核心計劃展覽中心位於汀九青山公路，利用模型、圖片、展板及錄影帶，介紹十項核心工程的資料。

機場核心計劃包括：

- 新機場
- 機場鐵路
- 青嶼幹線
- 西區海底隧道
- 北大嶼山快速公路
- 西九龍快速公路
- 西九龍填海計劃
- 中區填海計劃
- 三號幹線（青衣至葵涌）
- 東涌新市鎮發展

中心亦是供市民觀賞青馬大橋的最佳地點。

網址：www.info.gov.hk/napco/

注意事項：

- 地址：新界汀九青山公路四百零一號。(電話：2491 9202)
- 每天早上十時至下午五時，除公眾假期除外，逢星期一休息。免費入場。
- 聽從及遵守在場職員的指示。

交通： 於荃灣碼頭乘搭九巴234B路線往浪翠園或53路線往元朗東，於麗都灣站後下一站下車。
建議時間： 約2小時

工作紙：
– 參觀香港規劃及基建展覽館

1) 列出展覽館內有關下列方面的展品：

社會：＿＿＿＿＿＿＿＿＿＿＿＿＿＿＿＿

經濟：＿＿＿＿＿＿＿＿＿＿＿＿＿＿＿＿

環境：＿＿＿＿＿＿＿＿＿＿＿＿＿＿＿＿

2) 城市規劃—寫出維多利亞港的發展建議：

- ＿＿＿＿＿＿＿＿＿＿＿＿＿＿＿＿
- ＿＿＿＿＿＿＿＿＿＿＿＿＿＿＿＿
- ＿＿＿＿＿＿＿＿＿＿＿＿＿＿＿＿

3) 旅遊業—你認為香港的古蹟、文化建築物還是最新景點如迪士尼樂園，較能吸引遊客訪港？為什麼？

＿＿＿＿＿＿＿＿＿＿＿＿＿＿＿＿＿＿＿＿

＿＿＿＿＿＿＿＿＿＿＿＿＿＿＿＿＿＿＿＿

＿＿＿＿＿＿＿＿＿＿＿＿＿＿＿＿＿＿＿＿

4) 環境保護—空氣污染的原因及解決方法？

- ＿＿＿＿＿＿＿＿ 解決方法：＿＿＿＿＿＿＿＿
- ＿＿＿＿＿＿＿＿ 解決方法：＿＿＿＿＿＿＿＿
- ＿＿＿＿＿＿＿＿ 解決方法：＿＿＿＿＿＿＿＿

5) 基建走廊最吸引的特點是什麼？

＿＿＿＿＿＿＿＿＿＿＿＿＿＿＿＿＿＿＿＿

原因：＿＿＿＿＿＿＿＿＿＿＿＿＿＿＿＿

4.4 Applying Resources and Waste Management

With just a small area of 1,103 km^2, Hong Kong, surprisingly, has numerous natural landscapes such as forests, streams and reservoirs. These valuable resources require proper management for the use of local people. At the same time, the city also produces tremendous amount of waste, a problem that more people are becoming aware of thanks to the massive news reports about the landfill shortage.

Drinking water ➔

Landfills, reservoirs, refuse transfer stations and water treatment plants are perfect outdoor classrooms for understanding the principles of resources and waste management. In this chapter, students are encouraged to visit **landfill sites** of Environmental Protection Department (EPD), and **water treatment works** of Water Supplies Department (WSD).

1) EPD's Landfill Sites

Outlook of landfill site

Dumping waste

Thousands of tonnes of garbage from business, industry and residences are disposed of everyday. All managed by EPD, three existing strategic landfills namely West New Territories Landfill in Nim Wan, South East New Territories Landfill in Tseung Kwan O, and North East New Territories Landfill in Ta Kwu Ling, are the key disposal sites for Hong Kong.

These landfills have used advanced landfill technology in response to the more demanding environmental requirements and to provide the much-needed capacity for waste disposal. Liners, leachate collection and treatment systems, landfill gas management systems, and surface and ground water management systems are in place to control air and water impacts.

Among the three landfill sites, the **Southeast New Territories (SENT) Landfill** is recommended to visit. Located in the suburb of Tseung Kwan O, SENT landfill has an area of 100 hectares. It began operation in 1994 and is currently collecting 8,400 tonnes of domestic and industrial refuses per day. Visitors may learn about the latest technology as well as the problems of waste treatment in the landfills of Hong Kong.

Aerial photo of SENT

Notes on field visits:

- Make prior reservation with the contractor (i.e. Green Valley Landfill Ltd.) of SENT landfill.
- Put on suitable clothes and footwear.
- Do not collect any refuse or unknown objects from the landfill site.
- Listen and follow the advice of local management authority. You may also follow the **practical worksheet** at the end of this chapter.
- Pay attention to personal safety during the visit.

Transportation: There is no direct public transportation to SENT landfill. Visitors must arrange for their own transportation or travel by taxi to Wan Po Road inside Tseung Kwan O Industrial Estate.

Suggested time: 2-3 hours at the site.

Staff explaining waste treatment and operation of landfill to visitors

2) WSD's Water Treatment Works

Large tanks inside water treatment works of WSD

There are now 21 water treatment works in Hong Kong with a total output capacity of 4.8 million cubic metres of fresh water supply per day. During the water treatment process, raw (untreated) water is dosed with chemicals for mixing and then passed to the clarifiers where coagulation and flocculation in the water will take place. Settled water then flows into filters of sand or anthracite for removal of the more finely divided suspensions.

Sha Tin Water Treatment Works and **Pak Kong Water Treatment Works** have been opened to schools and groups for visit since September 1994. Staff of water treatment works will provide guidance and briefing during the visit.

Through the visits, students learn the knowledge on water supply in Hong Kong. Through introducing water treatment process, the principles of resources management and the responsibilities towards the environmental protection can best be demonstrated.

Staff explaining testing of water quality in the laboratory

Notes on field visits:

- Each visit group should consist of not more than 40 participants with at least 2 teachers or working staff.
- Participants shall arrange for their own transportation.
- Each guided visit will last for approximately 1 hour to 1 hour 30 minutes (from 9:30 a.m. to 11:00 a.m.)
- Listen and follow the advice of local management authority.
- Pay attention to personal safety during the visit.

Transportation: Take KMB Bus 88P from Shatin KCR station towards Hin Keng Estate. After arriving at the terminal, walk for 5 minutes and the Water Treatment Works is on the opposite side of the KCR railway.

Suggested time: about 2 hours at the premises.

Vising Water Treatment Works

Tanks of water to be treated

Guided tour provided

Practical Worksheet
- for visiting SENT (Tseung Kwan O) Landfill Site

1) What types of refuse do the landfill collect?

_________________ Treatment: _____________________________

_________________ Treatment: _____________________________

_________________ Treatment: _____________________________

2) Capacity of landfill and its expected lifespan:

_____________________ years left

Contingency plan: ____

3) Future use of site after landfill: ______________________________

4) What are the problems encountered by landfill:

- ____
- ____
- ____

5) Besides landfill and reclamation, are there any other waste treatment methods?

____

Adv : ____________________ Disadv: _________________________

6) Other questions/comments: ___________________________________

____

____

4.4 應用資源與廢物管理

香港面積雖小，佔地只有1,103平方公里，但卻擁有豐富的天然資源如森林和水塘，這些珍貴的資源必須恰當地管理才可供市民使用。另一方面，香港每天產生大量廢物，隨著回收箱日漸普及，加上有關堆填區迅速飽和的報導，市民漸漸已認識到此問題。

廢物回收箱

堆填區、水塘、廢物轉運站、濾水廠及污水處理廠都是學習資源和廢物管理的理想地點。我們建議參觀政府轄下的**堆填區**和**濾水廠**。

1) 堆填區

本港每天需要處置的工商業及住宅廢物數以千公噸計。目前，本港共有三個策略性堆填區，分別是位於稔灣的新界西堆填區、位於將軍澳的新界東南堆填區及位於打鼓嶺的新界東北堆填區。這些堆填區由環保署負責監督，是本港處置廢物的主要地點。

堆填區利用先進的堆填技術來應付社會愈來愈高的環保要求，並為本港提供處置廢物所需的地方。有關控制空氣及水質影響的措施包括防滲透墊層、滲漏污水收集及處理系統、堆填氣體管理系統、地面水及地下水管理系統。

香港現時共有三個堆填區，我們建議參觀位於將軍澳郊區的新界東南堆填區。這個堆填區於1994年開始運作，佔地100公頃，目前接收的家居及工業廢物約每日8,400公噸。訪客可透過參觀，認識香港堆填區所應用的最新科技及有關廢物處理的問題。

新界東南堆填區鳥瞰圖

注意事項:

- 參觀前須向有關承建商預約(如新界東南堆填區的承建商為翠谷工程有限公司)。
- 穿著適合衣服及鞋。
- 切勿在堆填區內拾去任何垃圾或不知名物件。
- 聽從及遵守在場職員的指示。你亦可參考本章的工作紙。
- 參觀時注意個人安全。

交通: 沒有公共交通前往新界東南堆填區。訪客須自行前往或乘搭的士到將軍澳工業村環保大道盡頭。

建議時間: 2–3小時

2) 濾水廠

香港現時共有21個濾水廠，每日可處理約480萬立方米淡水。在濾水過程中，原水(即未經處理的水)會先被混入化學品，進行攪拌，然後流入澄清池凝聚和沉澱，再流進濾水池，利用砂及無煙煤進行過濾。

沙田濾水廠及馬鞍山濾水廠 自一九九四年九月起開放給學校及團體參觀。濾水廠職員會帶領參觀者參觀及作出講解。

水務署歡迎各學校及團體參觀濾水廠，藉以提高市民對供水服務的認識。市民亦可透過認識濾水過程，了解水務署除了以提供可靠充足的優質食水為己任，亦會時刻關注保護環境須負的責任。

注意事項:

- 每組參觀人數的上限為40人，其中最少兩人必須為老師或工作人員。
- 參觀團體必須自行安排交通工具。
- 每次參觀活動為時約1小時至1小時30分(上午9:30至11:00)
- 聽從及遵守在場職員的指示。
- 參觀時注意個人安全。

交通: 從沙田火車站乘搭88P巴士往顯徑村，於總站下車步行約5分鐘，濾水廠位於火車站另一邊。

建議時間: 約2小時

工作紙：

– 參觀新界東南（將軍澳）堆填區

1) 堆填區主要接收什麼類型的垃圾?

______ 處理方法：______

______ 處理方法：______

______ 處理方法：______

2) 堆填區的總容量及預計壽命：______ 年

相應措施：______

3) 堆填區將來的用途：______

4) 堆填區面對的問題：

- ______
- ______
- ______

5) 除堆填和填海之外，其他處理垃圾的方法還包括：

______

好處：______ 壞處：______

6) 其他意見/建議：______

______

4.5 Applying Food Safety

Recent years have seen an increase in the number of food poisoning cases in Hong Kong, with more than 600 outbreaks on average each year. More than 2,000 people per year have been affected as a result, and about 60 per cent of cases have been traced back to catering premises. Although domestic food-poisoning outbreaks have also been recorded, some of these situations occurred after people ate take-away food. Food-poisoning cases are most prevalent in summer, and common items involved in outbreaks at food premises have included siu mei and lo mei dishes, sushi and sashimi, oysters and other seafood, hot-pot meals, salads, frozen confections and other ready-to-eat food. Consumption of food served in bulk, such as buffet and lunchbox meals, has also led to food poisoning.

According to the information provided by Food and Environmental Hygiene Department, there are several common and important types of food poisoning in Hong Kong, namely (see Appendix A for details):

1. Bacterial food poisoning,
2. Viral food poisoning,
3. Chemical food poisoning
4. Pesticide poisoning
5. Shellfish food poisoning
6. Ciguatera fish poisoning

Corn field

Examples of raw **genetically modified (GM) foods** are such as soybean, corn and tomato while processed GM ingredient containing foods are such as corn chip, potato chips, tofu, and soymilk etc. The Food and Environmental Hygiene Department points out that all GM foods that reach Hong Kong have undergone rigorous safety assessments by industry and regulatory agencies at the places of origin.

Safety checks for these products follow the "substantial equivalence" principle endorsed by WHO/FAO. The key considerations in determining whether a new food is substantially equivalent to an existing one include:

- Characteristics of the donor and the host organisms
- Composition
- Dietary intake
- Nutritional data
- Toxicological data
- Allergenic properties

Inside a typical supermarket

The first objective of this chapter is to learn how to record food poisoning outbreaks in Hong Kong using the Practical Worksheet on the next page as well as the information on Appendix A. The second activity for the students is to investigate food safety of supermarket products by accessing the contents of additives and preservatives as stated in the food labels. Based on the information on Appendix B and record your findings on the Practical Worksheet. At the end, give your conclusion on the safety of your selected products with reasons.

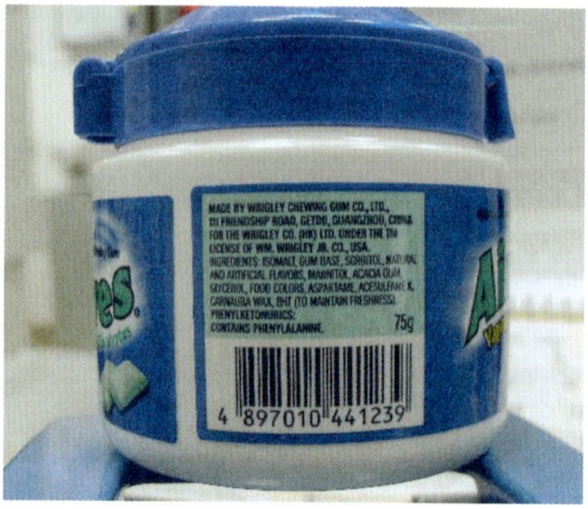

Ingredient list on processed food product

Notes on activity in supermarket:

- Go to a nearby supermarket (e.g. Park-N-shop or Welcome) in the afternoon during weekdays when there are relatively fewer customers shopping.
- Each visit should consist of at least 2 persons.
- Each visit will last for approximately 1 hour 30 minutes (e.g. from 2:30 to 4:00 pm).
- Bring a camera for taking photos of food products in the supermarket.
- Use the attached practical worksheet to record your findings.
- Don't violate the regulations of supermarket.

Suggested time: about 2 hours

Practical Worksheets

1. Food poisoning outbreaks

Please report the detail of a recent food poisoning outbreak in Hong Kong:

Types of food poisoning	
Date	
Venue	
Number of people affected	
Source (Food involved)	
Causative agent (e.g. Salmonella spp.)	
Onset period	
Symptoms	
Suggested Prevention	

2. Additives and preservatives in food products

Please find two different types of processed food in supermarket and count the number of additives in the products:

Food		
Manufacturer		
Acids, Bases and Buffers		
Antioxdiants		
Artificial Sweeteners		
Colouring Agents		
Emuslifiers, Stabilizers and Thickeners		
Flavour Enhancers		
Preservatives		
Others (please specify e.g. GM ingredients):		

Appendix A. Check List for Food Poisoning in Hong Kong (Extracted from Food and Environmental Hygiene Department)

1. Bacterial Food Poisoning

The number of cases for this type of food poisoning is the highest in Hong Kong. Different pathogens are come from different sources of food. *Vibrio parahaemolyticus* are come from raw or undercooked seafood, shellfish, marine products and salted food. *Salomonella* spp. are found in raw or undercooked egg and egg products (e.g. Tiramisu); undercooked meat, poultry and their products. *Staphylococcus aureus* are obtained from the foods which have been subject to a large amount of handling or with no subsequent cooking and reheating (e.g. lunch boxes and sandwiches); the food sources of *Bacillus cereus* are leftover rice, fried rice, meat products and vegetables. *Clostridium* spp., *Campylobacter* spp. and *Listeria monocytogenes* are among other pathogens found in contaminated food in Hong Kong. Most of them can be removed by proper storage of food and thorough cooking.

Raw or undercooked eggs are the main food sources for the pathogens – *Salmonella* spp.

Campylobacter spp. is mainly found in undercooked pork

2. Viral Food Poisoning

Norwalk-like viruses, the typical viral poisoning agents that come from raw or inadequately cooked shellfish especially filter feeders.

3. Chemical Food Poisoning

Clenbuterol is the veterinary drug in some contaminated pigs' offal or pork that would cause tachycardia, tremor, hypertension and muscle-relaxing effects.

4. Pesticide Poisoning

Pesticide poisoning is due to some pesticides remained in agricultural commodities such as leafy vegetables and fruits. There are several types of pesticides involved such as organophosphorus (e.g. methamidopos), carbamates, organochlorine and pyrethroids. Most of cases are due to the contamination of methamidophos.

Chinese kale (芥蘭)

Prohibited pesticides maybe found in leafy vegetables coming from mainland China such as Chinese flowering cabbage, Chinese white cabbage, spinach, water spinach, garland chrysanthemum, matrimony vinc and Chinese kale, etc. We can minimize the pesticides poisoning that wash and immerse vegetables and fruits in clean water for 1 hour or use boiling water to blench the vegetables and water used should be discarded.

Chinese flowering cabbage (菜心)

5. Shellfish Food Poisoning

Shellfish (such as clams, mussels, oysters, scallops) poisoning is due to consuming filter-feeding shellfish contaminated with phytotoxins produced by micro-algae. Some micro-algae are toxin-producing, the toxins will be concentrated in the filter-feeding shellfish tissue especially in viscera. After eating shellfish with phycotoxins, people can result in shellfish poisoning. There are four main types of shellfish poisoning - paralytic shellfish poisoning, diarrhoetic shellfish poisoning, neurotoxic shellfish poisoning and amnesic shellfish poisoning. So when you eat a smaller amount of shellfishes in one meal and avoid eating the viscera, gonad and roe.

Clams should be cooked thoroughly before eating.

6. Ciguatera Fish Poisoning

Ciguatera fish poisoning are reported increasingly and some of them are large outbreaks. It is due to the presence of ciguatoxin in the fish flesh and consumed by humans. The sources of ciguatoxin are coral reef and dinoflagellate that are found in coral reef. The ciguateric fish cannot be identified by the appearance. The larger the coral-reef fish and the higher the concentration of ciguatoxin present. The most common fishes involved are grouper, Russell's snapper red snapper, barracuda and moray eel. The prevention are to eat fewer coral-reef fish or eat small amount at any one meal; avoid eating the roe, liver, guts, head and skin of the coral-reef fish; avoid consuming alcoholic beverages simultaneously that would aggravate ciguatera poisoning.

Appendix B. Additives and Preservatives

An additive is added to food product either intentionally, to produce a desired effect or unintentionally through processing, storage or packaging. Preservatives are added to help maintain a food's freshness and keep it from spoiling, or oxidizing. Salt, sugar and vinegar were the first ways to preserve foods. There are some common food additives in the food products as follows. Some of them have harmful effects. If you find more additives in the processed food product, it will be more harmful to the body.

Basically the addictives include:

1. Preservatives
Benzoic acid and sodium benzoate – causes hyperactivity
Nitrites and nitrates of sodium and potassium – potentially carcinogenic.
Propionic acid and propionates
Sorbic acid and sorbates
Sulphur dioxide and sulphites – causes asthma and hyperactivity.

2. Acids, Bases and Buffers
Citric acid and lactic acid
Sodium hydrogencarbonate

3. Antioxdiants – that are used stopping the oxidation process
Ascorbic acid (Vitamin C) and ascorbates
BHA and BHT – may induce cancers
Propyl gallate
Sodium citrate
Tert-butyl hydroquinone

4. Artificial sweeteners
Acesulfame-K
Aspartame
Cyclamate
Saccharin

5. Emuslifiers, Stabilizers and thickeners
Carageenan – may cause allergies and intolerances
Carboxymethylcellulose
Detrins
Lecithins
Mannitol – causes allergies, diarrhea and nausea
Mono-glycerides and di-glycerides of stearic acid
Pectin
Potassium dihydrogencitrate
Sodium Alginate
Xanthan Gum

Food with addictives?

6. Flavour enhancers
Disodium 5' ribonucleotide – can cause skin rashes

7. Colouring agents
Tartrazine – causes asthma, hyperactivity, allergies and reacts with aspirin

4.5 應用食物安全

食物中毒事故的宗數近年在香港有上升的趨勢，平均每年發生600多宗食物中毒事故，逾2,000人受到影響，其中約百分之六十的食物中毒事故在食物業處所發生，部分家居食物中毒事故亦因當事人食用了從食物業處所購買的即食食物所引致。夏天是食物中毒爆發的高峰期，在食物業處所發生的食物中毒事故一般由以下食物引致：燒味、鹵味、壽司和刺生、蠔、海鮮、火鍋、沙律、冰凍甜點及其他即食食物，以及大量生產的食物，例如自助餐和午餐飯盒。

根據食物環境衛生署統計，下列為香港最常見和日漸普遍的食物中毒類別(詳見附件A)：

1. 細菌性食物中毒
2. 病毒性食物中毒
3. 化學性食物中毒
4. 農藥中毒
5. 貝介類食物中毒
6. 魚類雪茄毒中毒

未經煮熟的**基因改造食物**例子有大豆、粟米、蕃茄和馬鈴薯，而含有經過基因改造成分的食物例子則有粟米片、薯片、豆腐和豆奶等。根據食物環境衛生署的資料，所有基因改造食物必須通過該工業和來源地規管機構所制定的嚴格安全評估，才可在本港發售。

基因改造食物的安全評估，是根據世界衛生組織、聯合國糧食及農業組織和經濟合作與發展組織所認可的「實質等同」原則進行。評估會比較下列各項：

- 取出基因與接受基因的生物的特性
- 成分
- 攝取量
- 營養資料
- 毒性資料
- 過敏資料

本章第一項活動的目標是要學習如何記錄在香港發生的食物中毒事故，同學可利用下頁的工作紙及附件A的資料。第二項活動則是調查超級市場產品的食物安全，同學可閱參附件B的資料，再依據食物標籤所註明的添加劑和防腐劑份量來進行評估，請將調查結果記錄在工作紙，就所選擇的產品作出結論，並加以解釋。

在超級市場調查時應注意的事項：

- 於周日下午前往鄰近的超級市場(如百佳或惠康)，這時的購物人數會較少。
- 每次活動最少應有兩個人一同進行。
- 每次活動歷時大約個半小時(如2:30至4:00pm)。
- 攜帶相機以便拍攝超級市場的食品。
- 利用工作紙記錄調查所得的資料。
- 切勿違反超級市場的規則。

建議時間：逗留於超級市場約2小時

工作紙：

1. **食物中毒事故**

請記錄最近在香港發生的食物中毒事故的詳細資料：

食物中毒的類別	
日期	
地點	
受影響人士的數目	
源頭（牽涉的食物）	
病媒介（如沙門氏菌）	
病發期	
病徵	
預防方法	

2. **食品中的添加劑和防腐劑**

請比較兩種超級市場的加工食品的添加劑數目：

食品		
製造商		
酸、鹼及中和劑		
抗氧化劑		
人造糖		

食物色素		
乳化劑、穩定劑和增稠劑		
增味劑		
防腐劑		
其他（請註明如基因改造成分）:		

附件A. 食物中毒的列表（取自於食物及環境衛生署）

1. 細菌性食物中毒

香港最常見的食物中毒是細菌性食物中毒。不同的病原體所污染的食品也不同。生或未煮熟的海產、介貝類水產動物、海產製品和醃製食品會受腸炎弧菌污染。沙門氏菌會在生或未經煮熟的蛋及蛋類製品(例如意大利芝士蛋糕)及未經煮熟的肉類、家禽及其製品找到。金黃葡萄球菌則會污染需要很多處理程序的食品，但其後不經加熱或翻熱（例如飯盒和三文治）；仙人掌桿菌的食物源頭是剩餘的飯、炒飯、肉類製品及蔬菜。梭狀桿菌、彎曲桿菌和李斯特菌亦是香港常見的污染食物的病原體。預防細菌性食物中毒的最好方法是妥善貯存和徹底煮熟食物。

2. 病毒性食物中毒

諾沃克類病毒是常見的引致病毒性食物中毒的媒介，多附在在生或未徹底煮熟的介貝類水產動物。

3. 化學性食物中毒

克崙特羅是一種獸醫藥，可在某些受污染的豬內臟或豬肉找到，可引致心跳加速、震顫、高血壓及肌肉鬆弛現象。

4. 農藥中毒

農藥中毒是由殘留在葉菜和水果等農業產品的農藥所引致。常見的農藥包括有機磷酸鹽如甲胺磷、氨基甲酸酯、有機氯和除蟲菊酯。大部分的農藥中毒事故都是由甲胺磷引致。

從中國入口的葉菜可能含有禁用農藥，例子包括菜心，白菜菠菜，甕菜，塘蒿，枸杞和芥蘭等。為了預防農藥中毒，我們應將蔬菜和水果以清水洗淨及浸泡1小時或以沸水煮熟蔬菜，使用後的水應棄掉。

5. 介貝類食物中毒

介貝類(如蜆、蚌、蠔、帶子)食物中毒是由於進食了受藻毒素污染的濾食性介貝類水產動物。某些微型海藻會產生藻毒素，這些毒素會在濾食性介貝類水產動物組織尤其是內臟積聚。如果進食了受藻毒素污染的介貝類，便會引致介貝類食物中毒。介貝類食物中毒可分四類：麻痺性介貝類食物中毒、下痢性介貝類食物中毒、神經性介貝類食物中毒、失憶性介貝類食物中毒。每次只吃少量及避免進食介貝類的內臟、生殖腺和卵，可減低中毒機會。

6. 魚類雪茄毒中毒

魚類雪茄毒中毒事故近年有上升的趨勢，某些個案更屬於大型爆發。魚類雪茄毒中毒是由於人們進食了含有雪茄毒的魚肉所致。雪茄毒是來自生長於珊瑚礁和海藻的「雙邊毛藻」，我們並不能由魚的外表來辨斷該魚是否含有雪茄毒，但越大條海魚所含的毒素越高。常見含有毒素的海魚包括班、火點、紅鮪、鮫和鱔。要防止中雪茄毒，應減少進食珊瑚魚、每次只吃小量珊瑚魚；避免進食珊瑚魚的卵、肝、腸、魚頭和魚皮；避免同時喝酒致使加速雪茄中毒。

附件B. 添加劑和防腐劑

食品中加入添加劑可能是刻意為了製造理想的效果，但亦可能是加工、儲存或包裝時慎加入的。食品中加入防腐劑則有助保持食物新鮮及防止食物腐爛或氧化。鹽、糖和醋是保存食物的原始方法。下表列出常見的食物添加劑，某些可能對人體有害，若加工的食品含有添加劑的份量越多，對人體的傷害會越大。

添加劑基本上包括：

1. 防腐劑 苯甲酸和苯甲酸鈉 – 引致過度活躍 硝酸鈉和硝酸鉀的混合物 – 可能是致癌物質 丙酸 和 丙酸鹽 山梨酸 和 山梨酸酯 二氧化硫 和 二硫酸根 – 引致哮喘和過度活躍 **2. 酸、鹹及中和劑** 檸檬酸和乳酸 碳酸鈉 **3. 抗氧化劑 – 用作預防氧化作用** 抗壞血酸(維他命C)和抗壞血酯 丁化羥基茴香醚和丁化羥基甲苯 – 可引致癌症 棓酸丙酯 檸檬酸鈉 特丁基對苯二酚 **4. 人造糖** 醋磺內酯鉀 天冬胺、甜精及糖精	**5. 乳化劑、穩定劑和增稠劑** 紅藻膠 – 引致敏感 羧甲基纖維素 糊精 卵磷脂 甘露醇 – 引致敏感、腹瀉和作嘔 甘油和硬脂酸的甘油二酸酯 果膠 檸檬酸二氫鉀 海藻酸鈉 黃原膠 **6. 增味劑** 呈味核甘酸二鈉 – 引致紅疹 **7. 食物色素** 檸檬黃 – 引致哮喘、過度活躍、敏感及與阿斯匹靈產生作用

4.6 Applying Organic Farming

Organic farming is defined as a production system that avoids the use of chemical fertilizers, pesticides or genetically modified organisms. It is an approach to farming that emphasizes the importance of observing and co-operating with natural cycles, minimizing pollution and promoting sustainability. Ultimately, it aims to give a sustainable yield of healthy, good quality agriculture products without harming the environment.

Farmland

Examples of organic farming produce varieties include: supersweet corn, heat tolerant lettuce, coloured sweet pepper, spaghetti squash and zucchini, etc.

For group studying, students should visit **Kadoorie Farm and Botanical Garden** in Tai Po and **Produce Green Foundation Organic Farm** in Fanling.

Produce of organic farming

1) Kadoorie Farm and Botanical Garden (KFBG)

Farms at KFBG

Greenhouse

Kadoorie Farm and Botanic Garden nestles below Tai Mo Shan, Hong Kong's tallest mountain. Its deep-set valley has streams, woodland and terraces – together with conservation and education facilities. With 148 hectares, the Farm was established in the 1950s to bring agricultural aid to local poor farmers. With over 20 educational displays featuring themes of South China biodiversity, native wildlife conservation, organic farming and sustainable living, the Farm offers wide-ranging educational opportunities for public and receives over 140,000 visitors every year.

KFBG has its roots in agricultural practice. It embraces organic methods of farming and waste treatment, and other necessary components of sustainable agriculture. By setting a nurturing example for the community, the Farm promotes agricultural production systems that integrate human, environmental, and economic aspects in Hong Kong.

KFBG has practiced terrace farming for over 30 years. About 17 hectares of the land are dedicated to terraced orchards, growing over 25 different kinds of fruit crops. There was one incident in 1995: lychee and longan were unable to be harvested due to infestion by the Lychee Stinkbug. To tackle this problem, KFBG introduced a native parasitic wasp, *Anastatus japonicus*, to parasitize the stinkbugs' eggs. By the summer of 1997, the harvest had considerably improved, suggesting that this method is as effective as chemical sprays.

Juicy crops

Visitors can choose from a variety of wonderful products from the Farm. Every Sunday and public holiday, instant-grown organic produce is available for purchase in the reception area.

Organic produce

Allied with sustainable agriculture and creative nature education, KFBG plays an active role in promoting the conservation and biodiversity in Hong Kong and South China.

Website: www.kfbg.org.hk / **Enquiry**: 2488 1317

Poultry within the farm

Latest resident of KFBG

Transportation: Take KMB bus 64K towards Yuen Long (West) from Tai Po Market KCR station. Ride for 20-25 minutes and get off at the Kadoorie Farm stop.

Suggested time: 4-6 hours

Notes on field visits:

- KFBG opens from 9:30am - 5:00pm. No admission fee is required. KFBG also offers shuttle bus service for visitors on Sundays and public holidays. Fee for the ride is HK$20 per person.
- Put on suitable clothes and footwear.
- Do not damage or take away any objects.
- Follow the **practical worksheet** at the end of this chapter.
- As there are no restaurants in KFBG, visitors should prepare for their own meals.
- Brochures, maps, pot plant and souvenirs can also be found in the reception area.

2) Produce Green Foundation Organic Farm

Overview of organic farm

As a registered non-profit charitable organization, the Produce Green Foundation aims to promote organic farming, environmental friendly technology and green livelihood. The Foundation opened a farm in Hok Tau, Fanling in 1988. Since then, the farm has advocated the concepts of organic farming and environmental education to the public.

Farming activity

Bread-baking

Everyone can be a farmer!

Interested persons may visit and join the agricultural and cultural activities inside the farm. So far, more than 200,000 people have visited and many of them purchased local crops such as lettuce, beans, rice, carrot and other healthy products from the farm stalls.

Over 2,000 visitors have joined the courses to practice organic farming. The workshops include:

- Composting
- Natural Pest Control
- Chinese Herb Medicine
- Perennial Crop

Meanwhile, the Farm also provides free education packages to students and interested groups. Other interesting group activities include:

- Bread-baking
- Green handcraft (dying, knitting, paper handcraft)
- Green trip

Transportation: Take minibus towards "Hok Tau" outside the Fanling KCR Station. After about 20 minutes, get off at the roundabout of the top of the Lau Shui Heung Road. Then follow the signs and walk for a few minutes, go down the steps on the left, and you will arrive at the farm.

Suggested time: 3 hours at the farm

Website: www.producegreen.org.hk

Notes on field visits:

- You may arrange a group of 25 persons or above, by the name of an organization or by a person apply for group activities. Please make telephone reservation first (Tel: 2674 1190) and then provide written confirmation.
- Wear light and casual clothing. Bring more water in summer, and also food, hats and towels.
- Availability of farm produce depends on season while souvenirs and books are often available.
- Please take away all the stuff you bring into the farm, including garbage.

A totally refreshing experience!

Practical Worksheet
- for visiting Kadoorie Farm and Botanical Garden

1) Estimate the size of organic farm: ____________ hectares

2) Name some of the crops produced from the farm:

   Crop 1: _______________ Distinctive features: ______________________

   Crop 2: _______________ Distinctive features: ______________________

   Crop 3: _______________ Distinctive features: ______________________

   Crop 4: _______________ Distinctive features: ______________________

3) What are the advantages of organic farming, especially in KFBG?

   - ____
   - ____
   - ____

4) What are the drawbacks of organic farming?

   - ____
   - ____
   - ____

5) Will you buy products of organic farming?

   Yes / No, please give reasons: ______________________________________

   If yes, what would you buy: ____

6) Other comments/suggestions: ______________________________________

   ____

4.6 應用有機耕作

香港政府近年積極推廣有機耕種，這是一種絕不採用化學肥料、不用殺蟲劑或基因改造物的農業生產系統。有機耕種強調順應自然規律，減少污染及提倡可持續發展，旨在生產出健康而又高品質的農作物，並令地球資源得以永續。

有機耕種以尊重環境、重視生物多樣性及維持生態系統平衡為生產目標。現時廣受歡迎的新產品包括超甜粟米、耐熱生菜、彩色甜椒、魚翅瓜及翠玉瓜。

香港現時有很多進行有機耕作的農場，如要以團體形式參觀，我們建議到大埔的**嘉道理農場暨植物園**及粉嶺的**綠田園**。大家應謹記愛護大自然的守則—除了相片不帶走任何東西，除了腳印不留下任何東西。

1) 嘉道理農場暨植物園 (KFBG)

嘉道理農場暨植物園座落於新界中部的觀音山下，山谷內有溪流、林地和梯田。農場成於立1950年代，佔地148公頃，初時主要為貧窮的農民提供農業輔助。現時，園內設有環境保育與教育設施，包括20多個教育展覽，讓參觀者瞭解華南生物多樣性、本土動植物保育、有機耕作和可持續發展生活模式等保育專題，每年參觀人次逾14萬。

農場自成立以來一直以農務為本，奉行以有機方法耕作和處理廢料，這是可持續農業不可或缺的組成部份，透過給社會建立一個有機耕作的榜樣，農場在香港和華南地區推廣一個能兼顧人類生存、環境生態與經濟發展的農業生產模式。

嘉道理農場暨植物園有超過三十年的梯田耕種經驗，場內大約有十七公頃的梯田闢作果園，種植了超過二十五種果樹。一九九五年，荔枝和龍眼曾因荔枝蟲春象的侵擾而嚴重失收。農場於是引入本地的寄生蜂(*A. japonicus*)來對付荔枝蟲春象。到一九九七年夏天，收成明顯改善，這證明生物防治完全可以有效地代替化學農藥。

農場種植的有機蔬菜逢星期日及公眾假期會於接待處發售，遊客可向農場選購各類產品。隨著時代變更，嘉道理農場暨植物園近年積極在香港和華南地區推廣自然保育、生物多樣性、可持續耕作及自然教育。(網址：www.kfbg.org.hk)

注意事項：
- 農場的開放時間為上午九時三十分至下午五時。下午四時後謝絕遊客入場。入場費用全免。假日有穿梭巴士往來上下山區，每程收費$20。
- 穿著適合的衣服和鞋。
- 不可損毀或帶走任何物品。
- 聽從及遵守在場職員指示。你亦可參考本章的工作紙。
- 參觀嘉道理農場暨植物園至少需時半天，遊客需自備食物。
- 接待處備有地圖、資料單張、盆栽及名片套出售。(請致電2488 1317)

交通：於大埔墟火車站乘搭64K巴士，往元朗西方向，約20–25分鐘，在嘉道理農場暨植物園車站下車。

建議時間：約4–6小時

2) 綠田園基金

綠田園基金是一個註冊的非謀利慈善團體，旨在推廣有機耕種、環保科技及綠色生活。基金於1998年在粉嶺鶴藪成立農場，作為提倡有機耕種和環境教育的場地。

有機農場歡迎有興趣的人士前往參觀及參加他們所舉辦的耕種或文化活動，直至現時為至，已有超過200,000人到訪，他們大部分均有向農場購買本地的農作物如生菜、豆、稻米、蘿蔔及其他健康產品。

現時已有超過2,000位訪客曾參加有機耕種班，此外，農場亦成立多個工作坊，包括：

- 堆肥區
- 害蟲防治作物田
- 中草藥區
- 多年生的農作物區

基金亦會向學生和有興趣的團體提供免費的教育活動。

個人或團體活動的收費由$10至$98不等，其中包括：

- 有機耕種（半日工作坊）
- 綠色飲食日營
- 綠色手工藝（染料、編織和造紙）
- 生態漫遊

可參考網址：www.producegreen.org.hk

注意事項：

- 參觀人數如有25人或以上，請先致電農場預約團體參觀（2674 1190）及遞交書面申請。
- 請穿著輕便及合適衣服。
- 請帶備食物、毛巾和帽。夏天須帶多點水。穿著長袖衣服和長褲可避免蚊叮蟲咬。
- 訪客可向農場購買時令農產品、紀念品和書刊。
- 記謹帶走所有攜來的物品及垃圾。
- 聽從在場職員指示。
- 請不要取道鶴藪村進入農場，以免對村民造成滋擾。
- 綠田綠基金的經費來自市民及訪客的捐款，請慷慨解囊及支持！

交通：於粉嶺火車站乘搭開往鶴藪的專線小巴，車程約15至20分鐘，在流水響道迴旋處下車後按路牌指示沿車路步行約3分鐘，即見綠田園路牌在左邊，路牌旁邊有梯級，拾級而下便可到達農場。

建議時間：約3小時

工作紙：
- 參觀嘉道理農場暨植物園

1) 有機農場的面積：約 ______________ 公頃

2) 列出數種農場種植的的農作物：

農作物1：______________ 辨別特徵：______________________________

農作物2：______________ 辨別特徵：______________________________

農作物3：______________ 辨別特徵：______________________________

農作物4：______________ 辨別特徵：______________________________

3) 以嘉道理農場暨植物園來說，有機耕種的優點是什麼?

- ____
- ____
- ____

4) 有機耕種的的缺點是什麼?

- ____
- ____
- ____

5) 你會購買有機耕種的農產品?

會 / 不會，因為：____

若會，你會買什麼：____

6) 其他意見 / 建議：____

____

4.7 Applying Ecotourism

The concept of ecotourism has become a popular issue in Hong Kong especially since the outbreak of SARS in 2003. Generally speaking, ecotourism should:

- be nature-oriented;
- teach knowledge and conservation;
- promote local economy; and
- demonstrate sustainable development.

Ecotour at Mai Po Nature Reserve

Compared with ordinary tourism, ecotourism focuses on education and conservation. Thus, quality ecotourism rely on comprehensive planning such as environmental impact assessment.

Ecotourism should involve the following aspects:

- Geology (exploring natural landscape such as hills, streams, rocky shores, etc.),
- Ecology (exploring wildlife fauna and flora (birds, butterflies, dolphins, etc.), or
- Humanity (exploring human culture and history)

Ecotour Tsiu Hang Nature Trail,Sai Kung

Hong Kong has numerous natural locations for exploring the topic of ecotourism. Mai Po Nature Reserve, Nam Sang Wai, Tai Po Kau, Hoi Ha Wan are perfect examples to visit. Apart from these sites that were mentioned in previous chapters, it is also suggested to visit **Fung Yuen Butterfly Reserve** and **Ma Shi Chau Special Area** (both in Tai Po).

Remember the motto of ecotourism: "**take nothing but photos, leave nothing but footprints**".

1) Fung Yuen Butterfly Reserve

Located in the suburb of Tai Po, Fung Yuen is well known for its abundant butterfly diversity in Hong Kong and the East Asia region.

As early as 1980, Fung Yuen has already been named "site of special scientific interest"(SSSI). 8% (5 hectares) of the area belong to private property and has high conservation value. In 2004, the overall ecological richness of Fung Yuen ranked sixth throughout Hong Kong.

Common Tree Brown butterfly

The fantastic White Dragontail butterfly

According to the research conducted by Tai Po Environmental Association (TPEA) in 2003-04, there were over 150 butterfly species (about 60% of total species in Hong Kong) found in Fung Yuen. The common species include White Dragontail (燕鳳蝶), Common Tiger (虎斑蝶), Common Birdwing (裳鳳蝶), Peacock Pansy (美眼蛺蝶), Punchinello (波蜆蝶), etc.

There are three reasons why Fung Yuen possesses such high value of ecological diversity:

- Valley landscape (geological)
- *Feng shui* plant diversity (biological)
- Fruit farm and agricultural activities (humanity)

Managed by TPEA, Fung Yuen Butterfly Reserve (FYBR) has become a restricted area since 2004, with the aim to preserve the area and to educate the public on species conservation. For more details, refer to the website: http://www.fungyuen.org

Outlook of Fung Yuen Butterfly Reserve

Visitors especially butterfly-watchers come to Fung Yuen to enjoy the rich butterfly diversity and take photos of the species. Not only they won't hunt the species as specimens, they help in keeping the area clean and tidy (conservation) as well as spreading the news to other people (education).

Notes on field visits:

- Prior reservation should be made with TPEA for visiting FYBR (Tel: 27392481).
- Put on suitable clothes and hiking footwear.
- Do not take away any butterfly species or other insects from the reserve area.
- Listen and follow the advice of local management authority. Also follow the **practical worksheet** at the end of this chapter.
- Pay attention to personal safety during the visit.
- Take photos on the beautiful butterflies!!

Transportation: Take KMB Bus 75K from Tai Po Market KCR station towards Tai Mei Tuk. Take off at the first bus stop at Ting Kok Road. Then spend about 20 minutes to walk along the road path to the end. The FYBR Education Centre is located inside Mak Uk Village (No. 30).
Suggested time: 2 hours at the site.

Students visiting FYBR

Yellow Orange Tip butterfly

2) Ma Shi Chau Special Area

Visitors at Ma Shi Chau

Tombolo, a structure of sandy deposits connecting Ma Shi Chau and Yim Tin Tsai here. This geological feature is formed when sediment deposits built up and join from one island to another.

Ma Shi Chau Special Area comprises 4 islands in Tolo Harbour in Tai Po, namely, Ma Shi Chau, A Chau, Yeung Chau and an unnamed island. By walking through the Ma Shi Chau Nature Trail, one can understand the geology of the area with the on-site interpretation signs. Other than the geological features, fossils of ancient ammonites, bivalves, brachiopods, corals, moss animals, sea lilies and plants are also encountered.

Besides fossils, Ma Shi Chau is also an outdoor classroom to learn about plants and soil ecology. Lessons include the ages of the rocks, the fossils' colors and formation, as well as other exotic geological features include:

The oldest sedimentary rock - the oldest rock in Ma Shi Chau is located on its southeastern coast. It was formed during the Permian period (about 280 to 225 million years ago). The rock types are black shale, grey mudstone and siltstone.

"Dragons diving into the sea" - this geological feature is the result of differential erosion on a kind of sedimentary rock called shale.

As an ideal place for ecotours, the Ma Shi Chau Special Area is for all of us to protect and preserve. Excavation, collection and damaging of rocks, minerals and fossils within the special area are strictly prohibited.

Notes on field visits:

- Beware of the tide level before the trip and make sure you have adequate time. For tidal information, please call 187 8970 (24 hours) or 2926 8447 of Hong Kong Observatory.
- Leave the site before the tide comes in.
- Do not disturb the organisms and the ecology of the surrounding environment.
- Beware of sharp edges on rocks as well as loose, slippery and rugged ground surfaces.
- Keep the site clean and place rubbish in refuse bins.

Typical volcano rocks found in Ma Shi Chau

Transportation: Take Minibus No. 20K from Tai Po Market KCR station until the final station at Sam Mun Tsai. Then spend about 20 minutes to walk along the hiking path to reach Ma Shi Chau Special Area.

Suggested time: 2 hours at the site.

Practical Worksheet:
- for visiting Fung Yuen Butterfly Reserve

1) Area of Fung Yuen Butterfly Reserve: ____________ hectares

2) Name 4 butterfly species that you found in Fung Yuen:

_________________ Distinctive features: ______________________

_________________ Distinctive features: ______________________

_________________ Distinctive features: ______________________

_________________ Distinctive features: ______________________

3) What are the reasons that many butterflies can be found in Fung Yuen?

- ____
- ____
- ____

4) Problems encountered by / enemies of butterflies:

- ____
- ____
- ____

5) Please suggest some of the ways to conserve butterfly species:

- ____
- ____
- ____

4.7 應用生態旅遊

2003年沙士疫症在香港爆發後，生態旅遊的概念日漸普及。簡單來說，生態旅遊應包括以下元素：

- 以自然環境地點為主
- 具教育和環保意義
- 推動本土經濟
- 達到可持續發展

相對一般旅遊而言，生態旅遊更著重教育和環保意義，因此，優質的生態旅遊須視乎周詳的規劃如環境影響評估。

生態旅遊的主要項目包括：

- 地理考察（探索自然景觀如山坡、河溪、岩岸等。）
- 生態考察（探索野生動植物如雀鳥、蝴蝶、海豚等。）
- 人文活動（探索人類文化及歷史。）

謹記生態旅遊守則：**「除了相片不帶走任何東西，除了腳印不留下任何東西」**。

事實上，香港擁有豐富的自然資源可作生態旅遊。米埔自然護理區、南生圍、大埔滘和海下灣都是進行生態旅遊的理想地點，除了在各章提及的地點外，我們建議參觀位於大埔的**鳳園蝴蝶保育區**和**馬屎洲特別地區**。

1) 鳳園蝴蝶保育區

鳳園位於大埔區鄉郊，以其蝴蝶多樣性聞名於香港以至亞洲東區。早於1980年，鳳園已被列為「具特殊科學價值地點」，其中百分之八，或5公頃為私人土地及極具保育價值。鳳園的整體生態價值於2004年排行全港第六位。

根據大埔環保會(TPEA)於2003至2004年所作的調查，鳳園記錄得超過150種蝴蝶，佔全港蝴蝶品種超過六成。鳳園常見的品種包括燕鳳蝶、虎斑蝶、裳鳳蝶、美眼蛺蝶及波蜆蝶等。

大埔環保會於2004年成立鳳園蝴蝶保育區，進行保育工作及向公眾推廣物種保育的訊息。詳情請瀏覽 http://www.fungyuen.org

由於鳳園擁有豐富多樣的蝴蝶品種，因此一直是許多人尤其是觀蝶人士前往觀賞及拍攝蝴蝶的熱門地點，這些人士從不會帶走任何蝴蝶品種作標本，他們亦會保持當地清潔(保育)及向其他人傳播訊息(教育)。

鳳園擁有豐富的生態多樣性，其原因包括：

- 河谷地形（地理）
- 風水林的植物多樣性（生物）
- 果園及耕作活動（人文）

鳳園的農地將會進行大規模的發展計劃，居民已向發展商爭取保育自然生境，這是區內關乎可持續發展的重要事項。

注意事項：
- 鳳園蝴蝶保育區為受管制區域，由大埔環保會（TPEA）負責管理，訪客須預約才可參觀。
- 穿著合適衣服及行山鞋。
- 不可在保育區內捕捉蝴蝶或其他昆蟲。
- 聽從在場職員指示，你亦可參考本章的工作紙。
- 參觀時請注意個人安全。
- 記得為漂亮的蝴蝶拍照!!

交通： 從大埔墟火車站乘搭75K巴士往大尾督，於汀角路首個巴士站下車，沿鳳園路步行約20分鐘到達。

建議時間： 約2小時。

2) 馬屎洲特別地區

馬屎洲特別地區由吐露港上四個小島組成，包括馬屎洲、丫洲、洋洲及一個未命名的小島，這裡可讓你欣賞時間巨輪如何將地貌改變。馬屎洲自然教育徑更設有傳意牌，解釋各式各樣的地質特徵，此外，你更有機會發現化石如遠古的菊石、腕足動物、苔蘚蟲、珊瑚、海百合及植物。

除化石外，馬屎洲亦是學習植物及地理生態的戶外教室，遊人可在這裡學到岩石的年齡、化石的顏色和形成以及其他地理特徵如：

- **最古老的沉積岩** – 馬屎洲上最古老的岩石位於島上東南方岸邊。這些岩石在地質年代中的二疊紀形成，距今約二億八千萬至二億二千五百萬年。這些沉積岩多為黑色頁岩、灰色泥岩及粉砂岩。
- **"龍落水"** – 這條石龍是頁岩上出現「差別侵蝕」而產生的結果。

馬屎洲是進行地理考察的理想地點，這片地方需要你協助保護及保存。未經許可，任何人士不可在特別地區內挖掘、收集及破壞岩石、礦石及化石。參加者亦不可騷擾周圍環境的生態和生物。

注意事項：
- 考察前應查詢潮汐漲退的時間，並預留足夠時間供參加者於漲潮前離開。查詢潮汐漲退時間可致電香港天文台187 8001（24小時）或2926 8447（辦公時間）。
- 參加者須於漲潮前離開。
- 設計路線時應以各參加者都能完成為基本原則。
- 小心石塊的銳利邊緣、鬆散、濕滑及凹凸不平的地面。
- 保持場地清潔及將垃圾棄於垃圾桶內。
- 參加者須自行安排交通前往。

交通： 從大埔墟火車站乘搭20K小巴往三門仔，於總站下車，沿山徑步行約20分鐘到達馬屎洲特別地區。

建議時間： 約2小時

工作紙：

– 參觀鳳園蝴蝶保育區

1) 鳳園蝴蝶保育區的面積：________________ 公頃

2) 列出你在鳳園找到的四種蝴蝶：

________________ 辨別特徵：________________________________

________________ 辨別特徵：________________________________

________________ 辨別特徵：________________________________

________________ 辨別特徵：________________________________

3) 為什麼我們可在鳳園找到很多蝴蝶？

- ________________________________
- ________________________________
- ________________________________

4) 蝴蝶的天敵 / 遇到的問題：

- ________________________________
- ________________________________
- ________________________________

5) 請建議一些蝴蝶保育的方法：

- ________________________________
- ________________________________
- ________________________________

Reading Materials 讀本

Agriculture, Fisheries and Conservation Department (2001), *Knowing Corals*, Friends of the Country Parks and Cosmos Books Ltd., Hong Kong.

Agriculture, Fisheries and Conservation Department (2002), *Chinese White Dolphin*, Friends of the Country Parks, The University of Hong Kong and Cosmos Books Ltd., Hong Kong.

Agriculture, Fisheries and Conservation Department (2003), *101 Geographical Landmarks*, Friends of the Country Parks and Cosmos Books Ltd., Hong Kong.

Agriculture, Fisheries and Conservation Department (2003), *Knowing Hong Kong Marine Parks,* The University of Hong Kong and Cosmos Books Ltd., Hong Kong.

Agriculture, Fisheries and Conservation Department (2004), *Venturing Wetlands*, Friends of the Country Parks and Cosmos Books Ltd., Hong Kong.

Dudgeon, D. (2003), *Hong Kong Field Guides: Hillstreams*, The University of Hong Kong and Wan Li Book Co., Hong Kong.

Dudgeon, D. (2003), *Hong Kong Field Guides: Rocky Shores*, The University of Hong Kong and Wan Li Book Co., Hong Kong.

Hau, B.C.H. (2003), *Hong Kong Field Guides: Hillsides*, The University of Hong Kong and Wan Li Book Co., Hong Kong.

Hong Kong Bird Watching Society (2004), *My Favourite Hong Kong Bird*, The Hong Kong Bird Watching Society Limited, Hong Kong.

Hong Kong Discovery (2005), *Photographic Guide Series to Hong Kong Nature: Insect Portfolio – Hong Kong Insects (Volume 1 and 2)*, HK Discovery Limited, Hong Kong.

Leung, H.G.R and Chu W.L.M (2005), *Common Shrubland Plants,* Hong Kong Discovery Limited, ,Hong Kong

Williams, G.A. (2003), *Hong Kong Field Guides: Rocky Shores*, The University of Hong Kong and Wan Li Book Co., Hong Kong.

詹肇泰、邱榮光(2004)，*蝶、鳥、舞動——鳳園：具特殊科學價值地點*，香港：大埔環保協進會及萬里機構・萬里書店。

Useful Contacts 有用的聯絡方法

The following groups may be useful to know more about field trips and environmental studies in Hong Kong: 以下團體可提供更多有關野外研習和環境教育的資料：

Agriculture, Fisheries and Conservation Department 漁農自然護理署
Telephone / 電話：2422 9431
Website / 網址：http://www.afcd.gov.hk

China Lights and Power Group 中電集團
Website / 網址：http://www.clpgroup.com
Email / 電郵：csd@clp.com.hk

Environmental Protection Department 環境保護署
Website / 網址：http://www.epd.gov.hk
Email / 電郵：enquiry@epd.gov.hk

Food and Environmental Hygiene Department 食物及環境衛生署
Website / 網址：http://www.fehd.gov.hk
Email / 電郵：enquiries@fehd.gov.hk

Hong Kong Airport Core Programme 香港機場核心發展計劃
Website / 網址：http://www.info.gov.hk/napco/exhibition.html
Telephone / 電話：2491 9202

Hong Kong Bird Watching Society 香港觀鳥會
Website / 網址：http://www.hkbws.org.hk
Email / 電郵：hkbws@hkbws.org.hk

Hong Kong Discovery 野外動向
Website / 網址：http://www.hkdiscovery.com
Email / 電郵：info@hkdiscovery.com

Hong Kong Planning and Infrastructure Exhibition Gallery 香港規劃及基建展覽館
Website / 網址：http://www.info.gov.hk/infrastructuregallery
Telephone / 電話：3102 1042

Kadoorie Farm and Botanical Garden 嘉道理農場暨植物園
Website / 網址：http://www.kfbg.org.hk
Email / 電郵：info@kfbg.org

Produce Green Foundation 綠田園基金
Website / 網址：http://www.producegreen.org.hk
Email / 電郵：info@producegreen.org.hk

Tai Po Environmental Association 大埔環保協進會
Telephone / 電話：2739 2481
Website / 網址：http://www.taipoea.org

Water Supplies Department 水務署
Telephone / 電話：2824 5000
Website / 網址：http://www.wsd.gov.hk

Croucher Institute for Environmental Sciences, Hong Kong Baptist University
香港浸會大學 裘槎環科所

Telephone / 電話：3411 7746
Website / 網址：www.cieshk.org
Email / 電郵：cies@hkbu.edu.hk

A

Agricultural and Fisheries Conservation Department (AFCD) 24
Agriculture 8, 89–92
Animals – streams and shorelines 35–37
Animals – wetlands 10
Animals – woodlands 23
Aquatic 8, 36, 37

B

Birds – wetlands 8–14
Birds – woodlands 23
Butterflies and insects of woodlands 22

C

Cathay Pacific City 58
China Light and Power Company (ElectriCity) 59
Conservation – marine 38
Conservation – wetlands 10
Conservation – woodlands 24
Coral reef 8, 35, 39
Corporate environmental management 58
Country parks 24
Croucher Institute for Environmental Sciences (CIES) 1–2
Cultural heritage 8, 26

D

Diversity – streams and shorelines 34, 35, 39
Diversity – wetlands 8, 13
Diversity – woodlands 21, 24, 25

E

Economic 8, 10
Ecotourism 8, 96
Environmental awareness 51
Environmental Protection Department (EPD) 51
Environmental Resource Centres of EPD 51

F

Farming 8, 26, 89–92
Fish 8–13, 34–41
Food and Environmental Hygiene Department 79
Food poisoning 79–83
Food safety 79
Freshwater 8–11, 34–37, 41
Fung Yuen Butterfly Reserve 96

G

General guidelines for streams / shorelines field trips 41
General guidelines for wetland field trips 14
General guidelines for woodland field trips 27
Genetically modified (GM) food 79
Grasses and sedges 9

H

Hoi Ha Wan Marine Park 40
Hong Kong Airport Core Programme Exhibition Centre 67
Hong Kong Baptist University 1
Hong Kong Planning and Infra-structure Exhibition Gallery 65
Hong Kong Wetland Park 11

K

Kadoorie Farm and Botanical Garden (KFBG) 89
Key species – streams and shorelines 35
Key species – wetlands 8
Key species – woodlands 21

L

Lai Chi Wo 13
Landfill sites of EPD 72
Lions Nature Education Centre of AFCD 53
Lung Mun Country Trail 27

M

Ma Shi Chau Special Area 98
Mai Po Nature Reserve 12
Mangrove – wetlands 8–15
Mangrove – streams and shorelines 34, 35, 39, 40
Marine parks 38, 39
Marine reserves 38, 39

O

Organic Farming 89

P

Practical worksheet – applying corporate environmental management 61
Practical worksheet – applying environmental awareness 54
Practical worksheet – applying ecotourism 99
Practical worksheet – applying food safety 80
Practical worksheet – applying organic farming 92
Practical worksheet – applying resources and waste management 75
Practical worksheet – sustainable development 68
Practical worksheet – streams and shorelines 42
Practical worksheet – wetlands 15
Practical worksheet – woodlands 28
Produce Green Foundation Organic Farm 90

R

Ramsar Convention 11
Resources and waste management 72
Rocky shore 8, 34–39

S

Sha Lo Tung 26
Site of Specific Scientific Interest 26
Southeast New Territories Landfill (SENT) 72
Streams and shorelines 34
Supermarkets 80
Sustainable development 65

T

Tai Ho stream 41
Tai Po Kau Nature Reserve 25
Tidal zones 34
To Kwa Peng 14
Tung Ping Chau Marine Park 39

V

Vegetation – streams and shorelines 35, 37
Vegetation – wetlands 8
Vegetation – woodlands 22

W

Water Treatment Works of WSD 73
Wetlands 2, 7, 8–15
Woodlands 2, 7, 21–28

Note 記事

Note 記事

Note 記事

Note 記事

Note 記事